ADVANCE PRAISE FOR

Civil Society, Communication and Global Governance

"Global information and communication policies have been under public scrutiny as never before at the World Summit on the Information Society. This book provides an extremely timely and lucid exposition of what has taken place in this forum and what is at stake."
John D. H. Downing, Director, Global Media Research Center,
Southern Illinois University at Carbondale

"A must-read for all those interested in the encounters among United Nations politics, information issues and global civil society."
Cees J. Hamelink, Professor of International Communication,
University of Amsterdam

Civil Society, Communication and Global Governance

PETER LANG
New York • Washington, D.C./Baltimore • Bern
Frankfurt am Main • Berlin • Brussels • Vienna • Oxford

Marc Raboy & Normand Landry

Civil Society, Communication and Global Governance

Issues from the World Summit on the Information Society

PETER LANG
New York • Washington, D.C./Baltimore • Bern
Frankfurt am Main • Berlin • Brussels • Vienna • Oxford

Library of Congress Cataloging-in-Publication Data
Raboy, Marc.
Civil society, communication and global governance: issues from the World Summit
on the Information Society / Marc Raboy and Normand Landry.
Includes bibliographical references.
1. World Summit on the Information Society. 2. Information society—
Congresses. 3. Civil society—Congresses.
I. Landry, Normand. II. Title.
HM851.R33 303.48'33—dc22 2005017866
ISBN 0-8204-8112-2

Bibliographic information published by **Die Deutsche Bibliothek**.
Die Deutsche Bibliothek lists this publication in the "Deutsche
Nationalbibliografie"; detailed bibliographic data is available
on the Internet at http://dnb.ddb.de/.

Cover design by Lisa Dillon

The paper in this book meets the guidelines for permanence and durability
of the Committee on Production Guidelines for Book Longevity
of the Council of Library Resources.

© 2005 Peter Lang Publishing, Inc., New York
275 Seventh Avenue, 28th Floor, New York, NY 10001
www.peterlangusa.com

All rights reserved.
Reprint or reproduction, even partially, in all forms such as microfilm,
xerography, microfiche, microcard, and offset strictly prohibited.

Printed in Germany

Contents

List of Figures ix
List of Tables xi
List of Acronyms xiii
Foreword xvii
Acknowledgments xix

Part One: The WSIS in Context

Chapter One. The "Information Society": A Brief Introduction 3
   Regulation, Power and Security: The State in Question 3
   Information, Technology and Human Rights 5
   ICTs : Is it All About Money? 6
   Owning Knowledge: The Fight for Information 7
   Media Matters: Democracy in the Information Age 9
   The WSIS in Context 10

Chapter Two. Background to the WSIS 11
   Institutional Basis and Objectives 11
   Organizational Structures of the WSIS 14
   Functional Structures of the WSIS 17

Chapter Three. The Stakeholders 25
   The Actors 25
   NGOs at the United Nations 27
   Integration of Non-governmental Actors at the WSIS 29
   Competing Visions 32

Part Two: Civil Society at the WSIS

Chapter Four. Participating in the WSIS 39
    The Catalyzing Role of the CRIS Campaign 41
    The Role of UNESCO 43
    The African Regional Conference 44
    PrepCom 1 45

Chapter Five. The Organizational Structures of Civil Society 49
    The Civil Society Division 49
    The Civil Society Bureau 51
    The Civil Society Plenary in All its Stages 56
    On-line Communication 58

Chapter Six. Civil Society Demands at the WSIS 61
    The Battle for the Agenda 61
    The Main Obstacles 63
    Raising Consensus 65
    Civil Society's Content Development Efforts 71

Chapter Seven. Themes Raised by Civil Society 77
    Financing the Information Society 77
    Human Rights 79
    The Right to Communicate 82
    Intellectual Property Rights, Patents, Trademarks, and Public Domain ... 86
    Internet Governance 89
    Gender Issues 91
    The Media 93

Chapter Eight. Advancing Through the WSIS Preparatory Framework 97
    The Regional Conferences 97
    The Preparatory Committees 99
    The Summit 105
    Integration of Civil Society in a UN Meeting? 108

Part Three: Outcomes of the First Phase of the WSIS

Chapter Nine. The Official Outcomes of the WSIS 113

Chapter Ten. The Achievements of Civil Society.. 119
    Civil Society's Assessment of the WSIS Process.. 121

Chapter Eleven. The WSIS as a Model of Communication Governance 125
    The WSIS and Global Governance 126
    One Summit, Two Worlds.. 129

Appendix: "Shaping Information Societies for Human Needs" 135
Notes 163
Bibliography... 177

Figures

Figure 1. The ITU Vision of the WSIS. 13
Figure 2. Hierarchical Structures of the WSIS. 14
Figure 3. The WSIS Structure. 17
Figure 4. Civil Society's Participation in the WSIS. 55

Tables

Table 1. Organizational Structure of the WSIS 16
Table 2. Dates and Events of the First Phase of the WSIS 20
Table 3. Official Events and Outcomes of the First Phase of the WSIS 21
Table 4. Civil Society at the WSIS 54
Table 5. Civil Society Caucuses and Working Groups 57
Table 6. Documents Produced by Civil Society 74
Table 7. Intellectual Property Rights Controversies 87

Tables

Acronyms

ALAI:	Latin American Information Agency
ALER:	Asociación Latinoamericana de Educación Radiofónica
AMARC:	World Association of Community Radio Broadcasters
APC:	Association for Progressive Communication
BCUN:	Business Council for the United Nations
BIAC:	Business and Industry Advisory Committee to the OECD
CAMECO:	Catholic Media Council
CCBI:	Coordinating Committee on Business Interlocutors
CRIS:	Communication Rights in the Information Society
CSCG:	Civil Society Coordinating Group
CSB:	Civil Society Bureau
CSD:	Civil Society Division
CSP:	Civil Society Plenary
DOT Force:	G8 Digital Opportunities Task Force
ECOSOC:	United Nations Economic and Social Council
FAO:	Food and Agriculture Organization
FES:	Friedrich Ebert Stiftung
G8:	The Group of Eight
GDP:	Gross Domestic Product
GBDE:	Global Business Dialogue on Electronic Commerce
GIIC:	Global Information Infrastructure Commission
HLSOC:	High Level Summit Organizing Committee
IADB:	Inter-American Development Bank
IAEA:	International Atomic Energy Agency
ICANN:	Internet Corporation for Assigned Names and Numbers
ICAO:	International Civil Aviation Organization

ICC:	International Chamber of Commerce
ICTs:	Information and communication technologies
ILO:	International Labour Organization
IMO:	International Maritime Organization
IOM:	International Organization for Migration
IPR:	Intellectual property rights
IPS:	Inter Press Service
ITU:	International Telecommunication Union
MMI:	Money Matters Institute
MPAA:	Motion Picture Association of America
NGO:	Non-governmental organization
NWICO:	New World Information and Communication Order
OECD:	Organization for Economic Co-operation and Development
PrepCom:	Preparatory Committee
TRIPS:	Trade Related Aspects of Intellectual Property Rights
UDHR:	Universal Declaration of Human Rights
UN:	United Nations
UNCTAD:	United Nations Conference on Trade and Development
UNDP:	United Nations Development Programme
UNEP:	United Nations Environment Programme
UNESCO:	United Nations Educational, Scientific and Cultural Organization
UNFIP:	United Nations Fund for International Partnerships
UNFPA:	United Nations Population Fund
UNHCHR:	United Nations High Commissioner for Human Rights
UNHCR:	United Nations High Commissioner for Refugees
UN ICT TF:	United Nations Information and Communication Technologies Task Force
UNIDO:	United Nations Industrial Development Organization
UNITAR:	United Nations Institute for Training and Research
UNU:	United Nations University
UNV:	United Nations Volunteers
UPU:	Universal Postal Union
US:	United States

USCIB:	United States Council on International Business
WACC:	World Association for Christian Communication
WEF:	World Economic Forum
WFP:	World Food Programme
WGIG:	Working Group on Internet Governance
WHO:	World Health Organization
WIPO:	World Intellectual Property Organization
WITSA:	World Information Technology and Services Alliance
WMO:	World Meteorological Organization
WSIS:	World Summit on the Information Society
WSIS/ES:	World Summit on the Information Society Executive Secretariat
WTO:	World Trade Organization

Foreword

The subject of this document is an entirely new phenomenon in international politics: the active participation of non-governmental actors in the development of public policies at the supranational level. Since the Rio de Janeiro Earth Summit in 1992, more and more high-level international and multilateral meetings have been held on a variety of subjects. But "civil society"—a term used to refer to a vast and loose conglomeration of heterogeneous and more or less organized groupings and associations—had never been as integrated in such meetings as it was in Geneva in December 2003 during the first phase of the World Summit on the Information Society (WSIS).

This is a report on a meeting of two worlds. International diplomacy, embodied by the United Nations (UN), has evolved over a long history. The same can be said of civil society. And indeed, thousands of non-governmental organizations (NGOs) have had observer status with the UN and some of its affiliated agencies for many years. Since Seattle, other actors from all sorts of different horizons have demonstrated their interest in international public policy by making a considerable amount of noise at international meetings in venues as far afield as Washington, London, Prague and Quebec City.

However, the WSIS represents a profound break with these previous events because of the role played by civil society. Having been invited into the official process, the non-governmental actors clearly expressed their determination to be present at the centre of deliberations and to be considered as full partners. In this sense, the WSIS marked a shift from civil society's unrelenting challenges to the supranational decision-making process from the outside to its formal integration into just such a process on the inside. While civil society held firmly to its positions of principle on substantive issues of concern, it continued to be highly critical both of the way the process has unfolded and of its outcomes.

Civil society participants leveraged a conjuncture that was favourable to their inclusion in the WSIS negotiations to insist that the conditions of their participation be maximized. In the process they succeeded in speaking with a coherent voice—something that is, for obvious reasons, quite a feat in itself. Serious questions need to be raised as to the legitimacy of some of the interventions at the WSIS carried out under the civil society umbrella, just as similar questions can be asked about the legitimacy of many UN member States and the overall legitimacy of an international political process that excludes all of the world's populations from any direct involvement in its deliberations. The democratization of the emerging global system of political governance needs to be scrutinized from a variety of angles.

It is not by chance that this key moment in the evolution of the global governance system revolves around communication issues. Communication has rarely been in the forefront at major international meetings, but it is so today because it is the point of convergence of a series of key global problems and the possible solutions to them. Major UN events like the WSIS are arenas in which competing paradigms on the ways to address such issues are confronting each other and vying for positions as dominant models.

This means that the advocates of a different type of communication must wage their campaign on two quite opposite fronts, of citizen involvement on the one hand and politics on the other. The battle for public opinion can only be won through a broad process of public education and awareness-raising, which requires a strong presence at the national and international levels. At the same time, the political struggle directed at policy makers must necessarily channel the energy generated by educational work so that it can be used as a lever in the supranational arena.

The World Summit on the Information Society was an exceptional opportunity to put the emphasis on this longer-term effort. Our assessment is that the process has produced results, although it is far from over. The second phase of the WSIS, which at this time of writing was to take place in Tunis in November 2005 and devote its attention to digital solidarity and Internet governance against the backdrop of human rights, will fuel the political debate and exchange of ideas. Judging by the experience of WSIS, we can now rest assured that there will be a constructive, visible, watchful and effective presence of civil society in other forthcoming global debates.

Hence, this document, while seeking to report on a moment in the emergence of the new global communications environment, is also a snapshot of a historic "work in progress"—the emergence and self-assertion of civil society as a political actor on the world stage.

Montreal, May 2004 / August 2005

Acknowledgments

This book is a substantially revised and updated version of a research report originally written in French and published on-line in May 2004. Entitled *La communication au coeur de la gouvernance globale: enjeux et perspectives de la société civile au Sommet mondial de la société de l'information*, it was the first document to our knowledge that provided an exhaustive portrait as well as a critical analysis of the first phase of WSIS, the issues it raised and the role played by civil society. The report is still available on-line at:
www.lrpc.umontreal.ca/smsirapport.pdf

It was translated into English by Elvira Truglia and Hugh Ballem, with a grant provided by McGill University.

We would like to thank the *Fonds de recherche sur la société et la culture du Québec* and the Social Sciences and Humanities Research Council of Canada for the assistance they provided through their research support programmes.

We would also like to thank Jeremy Shtern, Isabelle Mailloux-Béïque and Daniela Bartosova for their valuable assistance during the research phase of this project.

We gratefully acknowledge permission of the Executive Secretariat of the World Summit on the Information Society for the use of copyrighted material appearing on pages 13, 14 and 17 of this book.

PART ONE
The WSIS in Context

The information society has its profits and its gurus. As a concept, the term is sufficiently blurred that it is prone to manipulation, following the wishes and needs of various actors. For some, it is the incarnation of a new global ideology, a vision of a globalized world that is unified and connected. For others, the information society refers to a new paradigm of communication, to the removal of barriers of exchange and to a final victory of technology over the constraints of space and time. For others still, the concept of the information society represents a rhetorical vehicle for a series of contradictory issues and interests. This global *politics* of information and communication brings together an ensemble of actors who find themselves in a situation of conflict and/or competition over strategic appropriation of the perceived gains of digital information and communication technologies.

A number of books have been devoted to critical discussion of the first two of these perspectives.[1] In the chapters that follow, we will principally focus on the third. More precisely, this book will study the perspectives and arguments put forward by the protagonists in a specific arena: the first phase of the World Summit on the Information Society (WSIS) which took place in Geneva in December of 2003. As a political framework, the WSIS has placed the governance of global communication on the world agenda, sparking a long overdue discussion that has, in turn, become the spearhead of a larger reconceptualization of the manner in which global decisions are made. The legacy of the WSIS will be more than just the impact of its political decisions; its true importance can only be appreciated through a reflection on the decisional processes themselves as well.

Our analysis will consist of three parts. The first part will focus on the context in which the WSIS emerged. We will then turn our attention towards the organization and structure of the Summit, in particular the place occupied by civil society (in terms of organization, participation and formal demands). In the third of this book's three sections, we will present a critical analysis of the first phase of the Summit.

Our study is clearly skewed by the emphasis we are placing on the role of civil society in the WSIS process. This is justified for several reasons. First, the WSIS marks a deviation from traditional practice in the manner in which global decision-making frameworks deal with non-governmental actors. Despite the fact that civil society actors did encounter serious problems and difficulties throughout their participation in the process, the WSIS must be seen as part of a larger trend towards the restructuring of global governance. Second, civil society participation at this Summit reflects the rising power of the voice of the citizenry at even the highest levels of international politics. Finally, and not least, the alternative discourses presented by civil society organizations throughout the WSIS process articulated a different conception of the information society which, when placed alongside the governmental positions, serves as a point of departure for evaluating the weaknesses, shortcomings, and biases of the official discourse.

• CHAPTER ONE •

The "Information Society": A Brief Introduction

The World Summit on the Information Society has its roots in a diverse, multi-disciplinary conceptual bouillabaisse. Since the beginning of the 1990's, we have witnessed a rising awareness of the transformative role played by information and communication technologies (ICTs) on a variety of dimensions of the human condition. Ever since, various actors have invoked the notion of an "information society" to represent a mutating social, economic and cultural reality which has been diametrically altered by these new technologies. The truth is that this phenomenon is eminently complex and refers to an ensemble of variables of very different natures. If the WSIS is an expression of a political will to engage this phenomenon, it must be approached with the goal of defining, grasping and understanding the information society.

In a modest way, we propose, in introducing the larger context of this book, to briefly review some of the dimensions that comprise the "information society" as well as certain issues surrounding the fundamental and still growing integration of ICTs into the heart of our way of life.

Regulation, Power and Security: The State in Question

The least that one can say is that traditional forms of media regulation are being put to a stern test by the development of new technologies. Satellite and Internet broadcasting make accessible contents and channels that may contravene the standards and regulations put in place by institutions charged with the enforcement of established norms at the national level. In regards to broadcasting, the basis for regulation lies in the recognition by all actors that the broadcasting frequency spectrum is a limited and scarce resource which must be allocated, fairly and transparently, according to an agreed upon understanding of the wider "public interest". In regards to digital broadcasting technologies, spectrum scarcity is neither an issue, nor a point of departure for regulation. In North America for example, certain radio presenters who have been evicted from the conventional airwaves for the shock jock nature of their material are now turning to satellite and digital radio, thus short circuiting the power of traditional regulatory institutions.

Legislation concerning the Internet is still in its embryonic stage. The tools that governments actually possess are not effective for regulating this network of networks. The truth is that the regulation of the Internet is a difficult task that necessitates a high level of legislative coordination between countries as well as a common political will to act. This common will often struggles to emerge when the interests of different States are brought into conflict with one another. Nonetheless, more and more States realize the necessity to take action in a number of areas which touch upon the regulation of the Internet, notably: in regards to accessibility, to tariffs, to fighting spam and to the hosting of hate speech sites as part of a larger battle against cybercriminality.

The emerging reality is that the range of activities covered by communication governance includes a great deal more than merely issues stemming from the use of communication technologies. Concerns over communication governance are increasingly integrated with concerns for two vital dimensions of the State: the exercise of governing and military defense. At a time when the rhetoric of "good governance" dominates political discourses, when criteria of transparency, of accountability and of dialogue serve as standards for challenging public authority, ICTs are being approached by many actors as tools of closing the gap between governments and the governed. It is currently in vogue to speak of "government on-line", of "e-democracy" and of forms of on-line consultations as though these technologies had a natural penchant for democracy. In fact, the reality is quite different and certainly less attractive. It is a slippery slope that begins with partnerships established between providers of technology and governments from authoritarian regimes. In order to reach new markets, certain multinational technology conglomerates have notably adapted their practices and their products to the conditions imposed by undemocratic countries. Microsoft, for example, has provided China with software that prohibits the use of words such as "democracy" and "liberty" in blogs. Tunisia, host of the second phase of the WSIS, is another example of a country whose government has drawn significant attention internationally for its repressive use of ICTs.

The US-led "war on terror" also raises very deep concerns about the gathering and arbitrary use of private and confidential data by public authorities. According to Amnesty International, the Patriot Act, adopted in haste by the

US Congress after the attacks on the World Trade Center, goes as far as allowing the US government to inquire on the reading habits of its people by monitoring libraries and bookstores. The US State now has the legal ability to conduct computer searches "*without providing notification, wiretapping and monitoring of e-mail*", as well as accessing private records.[1] As a counter-measure to terrorism, a whole international network of information gathering is now running very efficiently between the Western democracies. Obviously, authoritarian regimes do not have a monopoly on spying on citizens and arbitrary gathering of private information.

ICTs represent a major strategic issue in regards to defense and other military applications. Telecommunications centres are among the priority military targets of invading armies and every modern armed conflict includes a dimension of "information warfare" targeted at two fronts: global public opinion and pacification of hostile civilian populations. This phenomenon is nothing new; propaganda is as old as warfare. Nonetheless, today it is spreading to the new ICTs and taking form in different media configurations that we must study and understand.[2] In this regard, never have the resources necessary for the massive interception of private and public communication been as available to the institutions of foreign and domestic intelligence gathering as they are with the diffusion of digital ICTs and the Internet. This is not a reassuring development.

Information, Technology and Human Rights

No technology is neutral. The development of tools of communication always includes a political dimension. Because they spread ideas, values and opinions with an amazing efficiency and at relatively low cost, ICTs are at the centre of the oldest of democratic issues: the fight for the right to speak freely, for freedom of expression and of the press, for the protection of privacy, for participation in public debate and for the freedom from threat and oppression in pursuing these ends. This is true for "traditional technologies"—radio and television broadcasting as well as the press—as much as it is for those that define the contemporary digital age. In addition, ICTs raise questions about the protection of a whole series of rights in the information society: about the right to equitable access to technology and to knowledge; the right to the promotion and enjoyment of one's culture; the rights of women, children, the handicapped, indigenous communities and minority groups (be they based on ethnicity, religion or sexual orientation); as well as the right to protection against hate speech and racial discrimination.

At this time, none of these rights are secure—far from it. Internet users are imprisoned, sometimes for very long periods, for having published or even merely visited forbidden Websites. Intrusion, theft and unauthorized transmission of confidential private information are still common in cyberspace. In many cases the perpetrators of such violations have been the same parties who are ultimately responsible for the protection of human rights: representatives of the State. The fight against hate speech is also far from won. While many neo-Nazi Websites have recently been taken down in Europe, others have been put up on serves hosted outside the jurisdiction of European States. Despite recent developments, access to technology is still cruelly uneven. A communicational divide continues to separate men and women, rich and poor, young and old, educated and un-educated as well as urban and rural people. This gap is specific to neither the digital age, nor to recent technological development. Inequalities with respect to "old" technologies remain very significant. According to the United Nations Development Programme (UNDP), more than two billion people did not have access to electricity at the beginning of the twenty-first century. The ratio of telephone lines per capita drops from more than 1 for every 2 inhabitants in the rich countries of the Organization for Economic Co-operation and Development (OECD) to 1 for every 15 inhabitants in the developing countries and 1 for every 200 inhabitants in the least developed countries. This significantly limits the possibility for social and economic development in countries of the global South.

ICTs: Is it All About Money?

There is every indication that ICTs are acting as catalysts in the transformation of the economies of industrialized countries. ICTs also play an important role in accelerating the productivity growth needed to maintain a competitive advantage. OECD studies show that the ICT sector is growing steadily in terms of production, added value, jobs and trade. The intensity of the information and communication technology sector averaged 8.3% of GDP in OECD countries in 2001. Exports of ICT-related equipment accounted for up to 5% of GDP in some OECD countries.[3] Exports of ICT goods and services continue to grow at a significantly faster rate than GDP in these same countries. Pointing to analysis that predicts significant job creation and tremendous economic benefit, governments see the information society as a society of great promise for two reasons. First, as a marketplace, ICTs represent a highly competitive economic motor. The high added value of production in this industry drives forward the development of national economies. Furthermore, these technologies are contributing in varying degrees to a restructuring of post-industrial economies. This is initially felt at the level of employment. The increasingly prominent economic role of ICTs will force many countries to adopt measures to ensure that they can draw on a trained and skilled workforce. The sectors in which people are being hired and the criteria by which jobs are being staffed are being redefined in this way in conjunction with the penetration of digital technology. In addition, the structure of the corporation itself, the way in which business is done and the interaction between partners and competitors is being affected by these technologies. The collapse of the technology bubble has not prevented businesses from massively investing in cyberspace and from positioning information networks as part of integrated strategies which combine inventory and sales management, communication, marketing and the organization of work through information systems. It has become common to speak of "knowledge economies" in regards to countries where investment in ICTs has been high. Money is now being made without any concrete or material goods being produced. At the heart of this phenomenon we find an issue which will certainly be one of the most controversial discussions of the coming century: intellectual propriety rights (IPRs).

Owning Knowledge: The Fight for Information

In parallel to the development of a set of technologies and intangible assets, the countries at the cutting edge of research and development are staunchly defending their commercial interests at the national and international levels across various specialized institutions. Promotion of intellectual property rights is based on an often controversial regime of IPR laws which forms part of a complex system of international law. The debate over questions centering around IPR illustrates the profound differences between the actors involved and, indeed, a profoundly uneven world. In 1998, OECD member countries accounted for 86% of all patent applications and produced 85% of all technical and scientific articles. In the same year, the United States accounted for 54% of worldwide patent-related royalty and license fees, while a further 12% was earned in Japan.[4] The idea behind IPR—a concept which brings together copyrights, patents and trademarks—is that, to be stimulated and encouraged, innovation must remain under the control of its creators. In the absolute sense, the creator of an idea, an invention or a trademark is seen as being entitled to a limited monopoly over the innovation developed. As such, this logic follows, he or she should continue to be in charge of, and benefit financially from, its distribution and reproduction for a certain time. In fact, IPR regimes are extending the rights of copyright owners indefinitely, stifling innovation and creativity in the process. The spectrum of products and services covered by the regime of IPR at the international level is enormous. It includes the distributors of music and films, pharmaceutical companies, the creators and developers of software, public relations and marketing companies, authors in all genres, artists, editors as well as many other categories of actors who are less frequently associated with public discussions of IPR. According to the Motion Picture Association of American (MPAA)—a group which lobbies ferociously in favour of strengthening and enforcing the IPR regime—copyright industries brought in close to 90 billion dollars in exports and foreign sales to the US economy in 2001. Overall, the industry for copyrighted products comprises more than 5% of the gross domestic product (GDP) of the United States and employs more than 4.7 million people.[5] The MPAA suggests that the US motion picture business alone lost more than 3.5 billion dollars in 2003 from global piracy of its products. With the stakes being this high, it is not surprising that the economic issues in the debate over IPRs often deflect attention away from a series of social concerns that have also been raised in regards to the impact of this regime.[6]

Copyright pre-supposes that a certain level of financial capacity is required to cover the costs associated with creative production. Thus, copyright presumes one of two things about those who do not have both the means and the will to cover these associated costs: that they will be excluded entirely from access to copyrighted products, or, that if they do want or require such access to copyrighted products, they will obtain it either illegally or by contravening the copyright system. The critics of copyright and IPR in general believe that the regime, as it exists, is a veritable machine of social and economic exclusion, an instrument that principally serves the interests of wealthy corporations in the developed North to the detriment of users and creators everywhere and particularly those in the less developed South. Access to books, software, and inventions, not to mention knowledge and ideas is fundamental and, following this line of criticism, is limited, dictated and at the discretion of the rich industrial countries. Developing countries, led by Brazil and Argentina have recently been lobbying the World Intellectual Property Organization (WIPO) for the creation of a "development plan" that would re-establish more of a balance between the rights of producers and those of users. Though the WIPO is certainly one of the central battlegrounds in the confrontation, the impact of IPRs is sufficiently broad in its reach that similar debates are emerging in numerous other institutions of international governance.

Media Matters: Democracy in the Information Age

The links between the media and democracy have been well established. Only a free press, protected by a cadre of laws and clear constitutional rights, is capable of challenging public authority and of holding the different actors in society accountable to popular judgement. Challenges to this ideal will constitute one of the major democratic issues in the information society.

The last decades have seen the assertion of two major phenomena in regards to the media: an inflation of the mediated sphere—unprecedented development of channels and of new forms and contents of communication becoming accessible through ICTs—and an acute concentration in media ownership. Never before has so much information been produced or made available. This abundance is misleading in regards to the good health of our system of mass media, however. A handful of giant global corporations control increasingly large blocks of the world's media. Less than a dozen of them largely dominated the global media market at the end of the last century. Through vertical integration these companies create, produce and distribute their own products which they then are able to disseminate many times over, through their simultaneous holdings in numerous media forms such as magazines and newspapers, television and radio stations, films and on the Internet. This underlines a series of questions in regards to the priorities that shape the production and diffusion of news and entertainment content and the capacity of commercial media institutions to act in the public interest.[7]

A relatively recent form of activism positioning itself as a movement for media democratization is developing in strength and underlining the debate over a series of issues concerning the social role and impact of the media. The major issues pursued by this movement include questions of the role played by the media in our lives, the place and influence that advertising occupies in the shaping of media content, media ownership, citizen access to the media, propaganda and bias in the media, as well as the regulation of broadcasting. The media are, in this respect, an object of rising skepticism and are scrutinized by civil society with prudence and suspicion.

The WSIS in Context

As we have seen, the WSIS is taking place in a specific and eminently political context. Members of civil society, representatives of the private sector as well as international institutions and Heads of State met in Geneva in 2003 and again in Tunis in 2005 to promote agendas and positions that best serve their respective and common interests.

The issues surrounding the Summit are as numerous as they are complex. The following pages will focus on both the negotiation process and the issues at stake. By doing this, we hope to contribute to a better understanding of the idea of an "information society" and of the political space that will undoubtedly—at least in part—define it.

• CHAPTER TWO •

Background to the WSIS

> Developing countries should not forever be held hostage to the research agendas set by global market demand. If any form of development is empowering in the 21st century, it is development that unleashes human creativity and creates technological capacity. [...] Global initiatives that recognize this will not only provide solutions to immediate crises but also build means to cope with future ones.
> —UNDP, Human Development Report 2001 [1]

The decision to initiate the WSIS process was made at the 1998 ITU Plenipotentiary Conference (Resolution 73, Minneapolis), reiterated in 2000 (Resolution 1158 of the Plenary Meeting of the ITU) and formally stated in 2001 (Resolution 1179 of the Plenary Meeting of the ITU).[2] After discussion with the UN Secretary-General and the Administrative Committee on Coordination, the invitation was issued in resolution A/RES/56/183 of the United Nations General Assembly in December 2001, following a positive report on the Summit's feasibility.

Institutional Basis and Objectives

The United Nations General Assembly approved the decision to organize the World Summit on the Information Society on December 21, 2001 and delegated responsibility for leading the preparatory process to the International Telecommunication Union.

Resolution A/RES/56/183 is itself part of the framework set forth in the UN's Millennium Declaration and more specifically in Point 20, which states that the United Nations shall

> Ensure that the benefits of new technologies, especially information and communication technologies, in conformity with recommendations contained in the ECOSOC 2000 Ministerial Declaration, are available to all.[3]

The General Assembly of the United Nations also recognizes that

Information and communication technologies are among the critical determinants for creating a global knowledge-based economy, accelerating growth, raising competitiveness, promoting sustainable development, eradicating poverty and facilitating the integration of all countries in the global economy.[4]

The Okinawa Charter, adopted at the annual meeting of the Group of Eight (G8) governments in July 2000, also preceded and framed resolution A/RES/56/183 by identifying the main themes and issues that were to be taken up once the WSIS was structured. The discussions at the WSIS were to deal with these themes, focusing on a vision, access and applications:[5]

Vision: "To develop a common vision and understanding of the information society".

Access: "To promote the urgently needed access of all the world's inhabitants to information, knowledge and communication technologies for development".

Applications: "To harness the potential of knowledge and technology for promoting the goals of the United Nations Millennium Declaration".

The Vision of the ITU

The ITU Plenipotentiary Conference (Marrakech, 2002) identified three objectives for the WSIS:[6]

1. providing access to ICTs for all;
2. ICTs as a tool for economic and social development and meeting the Millennium Development Goals;
3. confidence and security in the use of ICTs.

Nine themes were officially adopted for the WSIS:

1. information and communication infrastructure: financing and investment, affordability, development and sustainability;
2. access to information and knowledge;
3. the role of governments, the business sector and civil society in the promotion of ICTs for development;
4. capacity building: human resources development, education and training;
5. security;
6. enabling environment;
7. promotion of development-oriented ICT applications for all;
8. cultural identity and linguistic diversity, local content and media development;

9. identifying and overcoming barriers to the achievement of the information society with a human perspective.

The ITU's initial documentation on the WSIS suggests a fundamentally technology-centred approach to the issues of the information society. The following figure encapsulates the UN agency's initial vision on the eve of the first meeting of the Preparatory Committee.

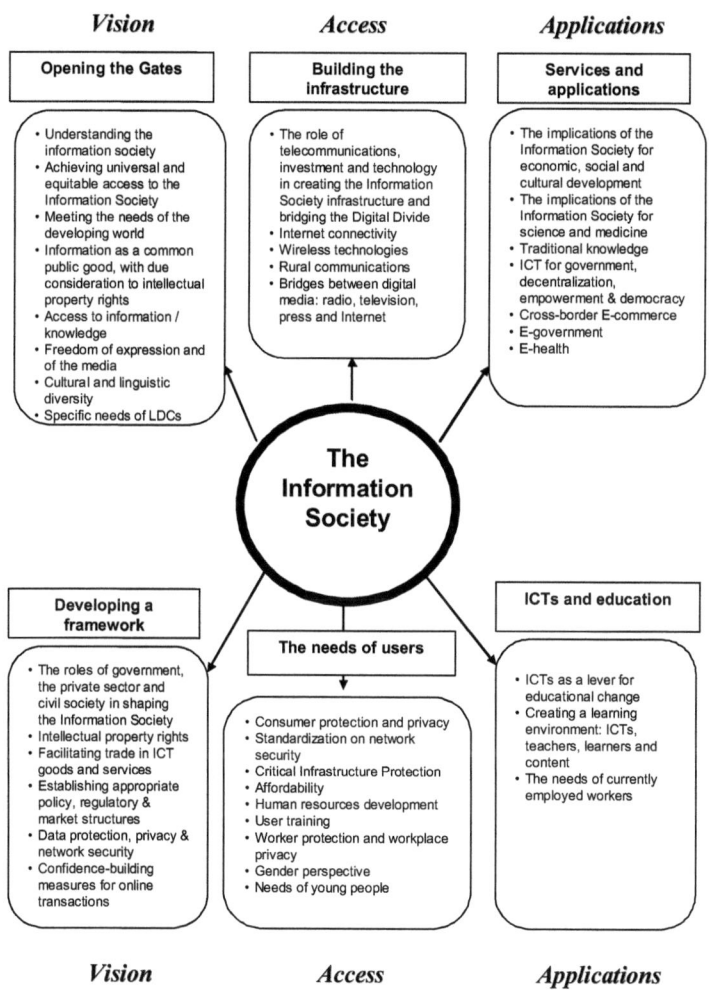

Figure 1. *The ITU Vision of the WSIS*[7]

The Organizational Structures of the WSIS

The WSIS was organized on the basis of a series of hierarchical structures of responsibility and jurisdiction. What was particular about the Summit was the formal involvement of a wide range of actors from UN agencies, governments, the private sector and civil society in its organization.

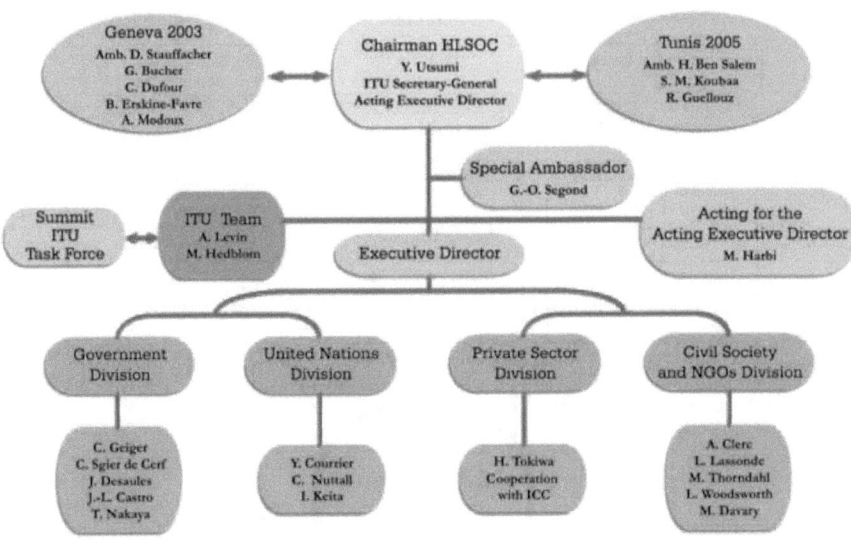

Figure 2. *Hierarchical Structures of the WSIS*[8]

The High Level Summit Organizing Committee (HLSOC)

Responsibility for organizing the Summit lay with a High Level Summit Organizing Committee officially established on March 11, 2001. The HLSOC's membership consisted of the executive directors of the specialized United Nations agencies concerned, namely FAO, IAEA, ICAO, ILO, IMO, ITU, UN Regional Economic Commissions, UNCTAD, UNDP, UNEP, UNESCO, UNFPA, UNHCHR, UNHCR, UNIDO, UNU, UPU, WFP, WHO, WIPO, WMO, World Bank, WTO. The following organizations also participated as observers: IADB, IOM, OECD, UNFIP, UNITAR, UNV. The Committee was chaired by ITU Secretary-General Yoshio Utsumi and was placed under the patronage of UN Secretary-General Kofi Annan. The Committee's job was mainly to coordinate the activities of the United Nations agencies in the preparation and organization of the Summit.

The Executive Secretariat (WSIS/ES)

The Executive Secretariat operated under the authority of the High Level Summit Organizing Committee to assist in the preparation of the WSIS. Pierre Gagné was its Executive Director until the end of the first phase. Two Host Country Secretariats were also created for the purposes of the Summit, one in Geneva for the 2003 phase, the other in Tunis for the second phase in 2005.

The Executive Secretariat's mandate was broken down into five main points:[9]

1. facilitate preparation of the Summit on behalf of all stakeholders;
2. prepare draft agendas and documents for Preparatory Committee meetings;
3. keep all stakeholders informed as to the evolution of the information society and preparations for the WSIS;
4. prepare progress reports on the preparation process;
5. secure sponsorships and funding for the preparation process and the Summit itself.

The Executive Secretariat was made up of representatives of the various parties involved in the Summit, i.e. governments, UN agencies participating in the HLSOC, the private sector and civil society. They were organized into four "divisions" responsible for liaising with the Summit's four partner groups, namely governments, UN agencies, the private sector and civil society.

The Swiss Secretariat

Switzerland, host country for the first phase of the WSIS, set up its own secretariat made up of individuals who were participating in the Summit as well as professional event organizers and managers. The head of the secretariat, Daniel Stauffacher, was also the Swiss ambassador to the WSIS.

Special Ambassador to the WSIS

During the first phase of the WSIS, a UN Special Ambassador to the WSIS was responsible for developing contacts with the high level representatives of governments and UN institutions to make them aware of the importance and the issues of the Summit. He was also responsible for contacts with the private sector and for making sure that the Summit received good media

16   • Civil Society, Communication and Global Governance •

coverage. The position was occupied by Guy-Olivier Second, former President of the State Council of the Republic and Canton of Geneva (Switzerland).

High Level Summit Organizing Committee (ITU)
Coordinates the actions of the United Nations agencies in the preparation of the Summit.

Executive Secretariat			
Governmental Division	UN Agencies Division	Private Sector Division	Civil Society Division
Each Division is responsible for external communications, works to enhance participation, promotes the diffusion of contributions and tries to make sure that the positions of its constituency are taken into account in the lead-up to the Summit.			

PrepComs and official structures

Bureaus		Organization of Procedures and Content Development		
Intergovernmental Bureau (created at PrepCom 1)	Civil Society Bureau (created at PrepCom 2)	Subcommittee 1 on Rules of Procedure (created at PrepCom 1)	Subcommittee 2 on Content and Themes (created at PrepCom 1)	Intergovernmental Working Groups (created at PrepCom 3)
The Bureaus are responsible for the logistical needs of their respective members.		Defines the rules and procedures for each kind of stakeholder.	Negotiates the content of the official texts with the governments.	Tries to find consensus at the intergovernmental level on particular thematic issues.

Autonomous Structures

Integrating Structures for the Private Sector and Civil Society	
Coordinating Committee of Business Interlocutors (CCBI)	Civil Society Plenary (created at PrepCom 1)
Bring together the contributions of their members in making consensual positions. Serve as spaces of exchange and discussion.	

Table 1. Organizational Structure of the WSIS[10]

Functional Structures of the WSIS

The first phase of the WSIS revolved mainly around two types of preparatory events: the Preparatory Committees (PrepComs), charged with moving the negotiations forward, and the regional conferences, which expressed the key priorities and policies of each of the world's large geographic regions. The predominance of governments over all other actors in these official activities was explicitly recognized. Non-State organizations were confined to the status of observers and advisors.

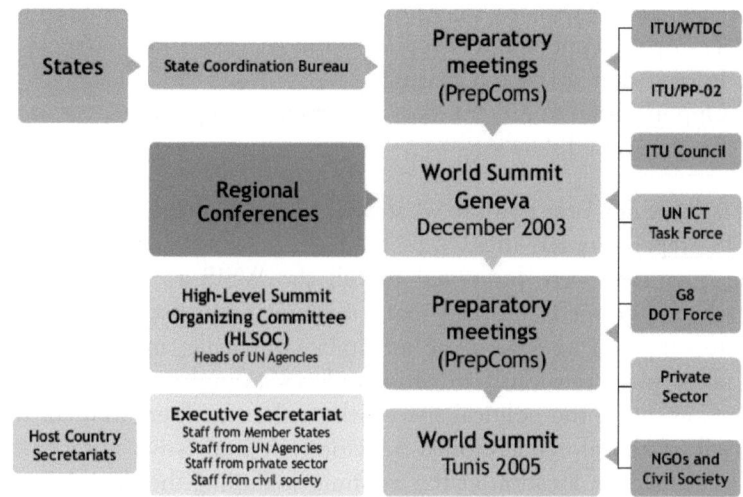

Figure 3. *The WSIS Structure*[11]

Given that their responsibility was to ensure the material organization of the Summit and to liaise with the main partners, the HLSOC and the Executive Secretariat were not supposed to play a primary role in developing the content of the WSIS. This was the prerogative of the participants themselves as they voiced their points of view at the Summit's subsidiary events, particularly the regional conferences and the preparatory meetings. The preceding figure, from an official document published at the very beginning of the WSIS preparatory process, shows the path that inputs to the Summit were supposed to take in theory. There are a number of points worth making about this figure: (1) first of all, one is struck by the predominant role assigned to States compared to the other main categories of actors, with UN agencies not even appearing as sources of input; (2) a key role is curiously bestowed on the G8 Dot Force, set up at the annual meeting of the G8 Heads of State at Kyushu-

Okinawa in July 2000; while the choice of the Dot Force as the G8 countries' means of intervening at the WSIS was understandable, one wonders by what right it enjoyed the status of a key source of input; (3) the figure ascribes a great deal of importance to the regional conferences, which are placed on the same footing as the preparatory meetings; yet, as we will see later, the final colouring of the Summit's outcomes was almost exclusively determined by the PrepComs. The WSIS's structures actually evolved to a significant extent as the preparatory process moved forward.

The official documentation of the WSIS declared the following principles to be fundamental to the Summit:[12]

- Transparency at all levels
- Information access and sharing
- Opportunities to present ideas, arguments and positions
- Openness in discussion

We will see at a later point to what extent these principles were reflected in the WSIS preparatory activities.

The organizers were prompted to call the WSIS a *"Sommet marguerite"* ("daisy wheel Summit"), meaning that the main inputs used to draft its final texts were supposed to come from a number of smaller international events having taken place previously in various places around the world. In reality, the PrepComs were the political spaces where the texts were drafted. In addition to all of the afore-mentioned activities, an impressive number of side events were held on location at the Summit, including the World Electronic Media Forum, the World Forum on Communication Rights and the World Forum on Community Media, as well as the ICT For Development Platform exhibition and conferences. The impact of some of these events will be discussed in the second part of this document.

Preparatory Committees

The PrepComs were essential pillars of the WSIS's functional structure. Their purpose was to define the themes and directions for the subjects to be discussed in order to reach a consensus on a *Plan of Action* and a *Declaration of Principles* at the WSIS. The PrepComs also set the procedures, and were therefore the arenas in which all of the stakeholders came together for the main negotiations. The status of the PrepComs was determined on the basis of Article 2 of UN Declaration A/RES/56/183 on the World Summit on the Information Society, which

Recommends that the preparations for the Summit take place through an open-ended intergovernmental Preparatory Committee, which would define the agenda of the Summit, finalize both the *Draft Declaration* and the *Draft Plan of Action*, and decide on the modalities of the participation of other stakeholders in the Summit.[13]

Declaration A/RES/56/183 also favoured the development of a participatory multi-stakeholder platform involving various UN agencies, the private sector and civil society. The Executive Secretariat was given the mandate to *"prepare draft agendas and draft documents for the PrepComs"*,[14] intended to elicit contributions from the various actors involved. Theoretically, the draft agenda was worked out through active consultations with the governments. Formal and informal meetings were one of the main ways these thematic or regional consultations were conducted. The contributions of the various actors present at the PrepComs were to be reported on in a document subject to a vote of approval in accordance with the WSIS Rules of Procedure.

The accreditation of non-governmental entities was a prerequisite for their participation in the PrepComs. The Executive Secretariat evaluated each accreditation application on the basis of the correlation between the applicant's activities, the applicant's involvement in the information society, and the theme of the Summit. The Secretariat subsequently made its recommendation on the issuance of the accreditation.[15] The forms of participation granted to non-State organizations were adapted in light of previous experiences at UN meetings.

The participation of non-governmental parties at the WSIS was prescribed by the explicit rules adopted at PrepCom 1 and set forth in the *Report by the Chairman of Subcommittee 1 on Rules of Procedure* (WSIS/PC-1/DOC/0009).[16] Under these rules all actors other than member States of the United Nations or its agencies are excluded from the negotiations, further confirming their exclusively advisory role at the WSIS.

The UN agencies, private sector entities and members of civil society would therefore take part in the preparatory meetings as observers, without the right to vote on proposals.[17] As we shall see, PrepCom 1 thus marked the burying of the hopes of civil society for a truly open and participatory Summit.

The contributions of the various actors were fundamental to the Summit's preparatory process. Their significance was even greater for actors who did not have voting rights and for whom the preparatory process was the main means of influencing decision-making at the WSIS. The rules governing inputs therefore defined, to a great extent, what actions were possible within the framework of the Summit.

Adama Samassékou, President of the *Académie africaine des langues* and former Education Minister of Mali, was elected President of the Summit Preparatory Committee for Phase I at PrepCom 1. The functions of the President were highly political, as he was responsible for the good course of the negotiations between stakeholders.

The Regional Conferences

The regional meetings were intended to design and define the specific needs and expectations of the different actors in each region with respect to the information society. In addition to the governments concerned, these meetings drew members of civil society and the private sector involved in each region as well as a number of foreign States with particular interests in the region. Beyond producing a declaration that served as an official input to the Summit, each of these meetings was an opportunity for the region's actors to hold conferences, round tables and seminars and to form working groups to exchange ideas and develop, through alliances and cooperation, action plans to address local needs.

Global Preparatory Meetings	*Regional Meetings*	*WSIS-related Conferences*
• PrepCom 1, Geneva, 1–5 July 2002 • Informal meeting on contents and themes, Geneva, 16–18 September 2002 • PrepCom 2, Geneva, 17–28 February 2003 • Intersessional Meeting, Paris, 15–18 July 2003 • PrepCom 3, Geneva, 15–26 September 2003 • PrepCom A, Geneva, 10–14 November 2003 • PrepCom 3B, Geneva, 5,6,9 December 2003	• Africa, Bamako, 28–30 May 2002 • Pan-European, Bucharest, 7–9 November 2002 • Asia-Pacific, Tokyo, 13–15 January 2003 • Latin America and the Caribbean, Bavaro, 29–31 January 2003 • Middle East, Beirut, 4–6 February 2003	• Bishkek-Moscow Sub Regional Conference, Bishkek, 9–12 September 2002, and Moscow, 23–24 October 2002 • Mauritius Conference on Access to ICTs by All, Pointe aux Piments, 3–5 April 2003 • 1st Conference of Ministers on Information & Broadcasting in the Asia-Pacific region, Bangkok, 27–28 May 2003 • Pan-Arab Regional Conference on the WSIS, Cairo, 6–18 June 2003 • World Information Technology Forum, Vilnius, 27–29 August 2003 • World Summit of Cities and Local Authorities on the Information Society, Lyon, 4–5 December 2003 • The Role of Science in the Information Society, Geneva, 8–9 December 2003

Table 2. Dates and Events of the First Phase of the WSIS

Objectives of the Global and Regional Meetings

One of the functions of the Executive Secretariat was to ensure that the contributions of the actors participating in the various conferences were comprehensively merged with the contributions from PrepComs and regional meetings in consensus documents that would serve as the basis for the *Declaration of Principles* and *Plan of Action* of the WSIS. The negotiating process that took place at these events was at the very heart of the emerging political understanding of the information society. Getting specific content into the texts was thus a strategic imperative for all involved in the process.

The retention and rejection procedures used by the Executive Secretariat to perform this duty were unclear. The Secretariat had a certain discretionary power when it came to drafting the documents for negotiation.

Official Negotiations and Advances Toward the Declaration of Principles and the Plan of Action		
Official Events	Documents Produced	Contents of Documents and Impacts on the Process
28-30 May 2002: Regional WSIS preparatory meeting in Bamako	*Bamako Declaration*	Principles and actions that the African region supports at the WSIS. Inputs into the drafting of the official WSIS *Declaration of Principles* and *Plan of Action*
1-5 July 2002: PrepCom 1	Internal regulations	Determination of the forms and conditions of participation of the various categories of actors and of the Summit's procedures
16-18 September 2002: Informal meeting on content and themes	Informal thematic discussion documents	Informal input on the themes to be discussed and directions for the texts to be adopted at the WSIS
7-9 November 2002: Regional pan-European WSIS preparatory meeting in Bucharest	*Bucharest Declaration*	Principles and actions that the European and North American region supports at the WSIS. Inputs into the drafting of the official WSIS *Declaration of Principles* and *Plan of Action*
13-15 January 2003: Regional WSIS preparatory meeting of the Asia-Pacific region in Tokyo	*Tokyo Declaration*	Principles and actions that the Asia region supports at the WSIS. Inputs into the drafting of the official WSIS *Declaration of Principles* and *Plan of Action*

Continued on next page

Table 3—Continued

29-31 January 2003: Regional WSIS preparatory meeting for Latin America and the Caribbean in Bavaro	Bavaro Declaration	Principles and actions that the Latin American region supports at the WSIS. Inputs into the drafting of the official WSIS *Declaration of Principles* and *Plan of Action*
4-6 February 2003: Regional Western Asia WSIS preparatory meeting in Beirut	Beirut Declaration	Principles and actions that the Western Asia region supports at the WSIS. Inputs into the official WSIS *Declaration of Principles* and *Plan of Action*
17-28 February 2003: PrepCom 2	No official document of significance was produced	WSIS starts consolidating inputs in order to draft the *Declaration of Principles* and *Plan of Action*
15-18 July 2003: Intersessional Meeting	*Draft Declaration of Principles* and *Plan of Action*	Official negotiations on the wording of the final documents of the Summit
15-26 September 2003: PrepCom 3	*Draft Declaration of Principles* and *Plan of Action*	Official negotiations on the wording of the final documents of the Summit
10-14 November 2003: PrepCom 3A	*Draft Declaration of Principles* and *Plan of Action*	Official negotiations on the wording of the final documents of the Summit
5,6 & 9 December 2003: PrepCom 3B	*Draft Declaration of Principles* and *Plan of Action*	Official negotiations on the wording of the final documents of the Summit
10-12 December 2003: WSIS	*Draft Declaration of Principles* and *Plan of Action*	Final texts of the first phase of the WSIS: adoption of the *Declaration of Principles* and *Plan of Action*

Table 3. Official Events and Outcomes of the First Phase of the WSIS

The Subcommittees

Two Subcommittees were set up as part of the preparatory process to assist the Preparatory Committee in its work. Subcommittee 1 on Rules of Procedure was tasked with defining and setting the rules under which the Summit would operate, in particular by determining the role and conditions of participation of each player in the process. The mandate of Subcommittee 2 on Content and Themes was to gather the various government contributions together and produce thematic documents reflecting the evolution of the negotiations. The documents so produced would then be submitted for adoption by the member States at the PrepComs. The conditions for participation in the Subcommittees were particularly important for the various stakeholders because they determined the balance of decision-making power in setting the agenda of the preparatory process. The rules on inputs to the Summit were stated in the report of the Chairman of Subcommittee 1 at the first PrepCom:

> Report by the Chairman of Subcommittee 1 on Rules of Procedure[18]
>
> Rule 44
>
> Subcommittees
>
> 1. The Preparatory Committee may establish subcommittees as it deems necessary for the performance of its functions.
>
> 2. Except as otherwise provided in these rules, each State participating in the Preparatory Committee as well as the European Community may be represented on each subcommittee.
>
> Rule 55
>
> Representatives of non-governmental organizations, civil society and business sector entities
>
> 1. Non-governmental organizations, civil society and business sector entities accredited to participate in the Committee may designate representatives to sit as observers at public meetings of the Preparatory Committee and its subcommittees.
>
> 2. Upon the invitation of the presiding officer of the body concerned and subject to the approval of that body, such observers may make oral statements on questions in which they have special competence. If the number of requests to speak is too large, the non-governmental organizations, civil society and business sector entities shall be requested to form themselves into constituencies, such constituencies to speak through spokespersons.

The UN agencies, regional commissions, international organizations, the private sector and civil society were excluded *de facto* from the debates and from participating in the Subcommittees, although they were allowed to attend. Disagreement on this rule not only dominated the discussion, but also sparked much heated exchange at PrepCom 1.

The ITU Council Working Group on WSIS

The ITU Council set up a working group on the WSIS, open to all UN member States and to members of the ITU sectors (private businesses, national and international organizations) interested in working on the ITU's contribution to the Summit's Preparatory Committee.

The Digital Opportunity Task Force (DOT Force)

At the Kyushu-Okinawa Summit in July 2000, the G8 Heads of State agreed to give closer consideration to the issues of the information society and the digital divide. They adopted the Okinawa Charter calling for the creation of a Digital Opportunity Task Force (DOT Force) made up of members of the public, the private sector and not-for-profit organizations. The DOT Force provides expertise to G8 members and develops plans of action in response to issues relating to telecommunications and the information society.

The ICT Task Force

The ICT Task Force was set up in March 2001 by the Secretary-General of the United Nations at the request of the Economic and Social Council in order to *"bring a truly global dimension to the many efforts underway to bridge the digital divide and take up the challenge of the digital opportunity, and thereby put ICTs in the service of development for all"*. The ICT Task Force has been asked to contribute to WSIS preparations by providing inputs and expert opinions.

These were the main official functional structures of the WSIS. We now move to a critical examination of the role of the respective broad categories of stakeholders.

• CHAPTER THREE •

The Stakeholders

> And to this day, Rio has become the benchmark against which future conferences and summits are measured in terms of civil society response—whether it be the summit on women in Beijing, the human rights conference in Vienna, Habitat in Istanbul, the population conference in Cairo or the conference on climate change at Kyoto last December.
>
> —Kofi Annan (1998)[1]

The Actors

The ultimate ends of the WSIS were determined both by the needs expressed by the participants and, to a large extent, by the relative positions of strength which underlay their relationships. These positions of strength related to the *nature* of the various participants and to their *positions* at the WSIS.

States have the power to legislate and to determine, through their economic and social development policies, how much consideration will be given to the development of the information society. The principle of national sovereignty ensures the predominance of States at the United Nations, whose specific culture in this respect is reflected most notably in the procedural provisions of Article 2 of the Declaration on the World Summit on the Information Society (A/RES/56/183):

> (The General Assembly) recommends that the preparations for the Summit take place through an open-ended intergovernmental Preparatory Committee, which would define the agenda of the Summit, finalize both the *Draft Declaration* and the *Draft Plan of Action*, and decide on the modalities of the participation of other stakeholders in the Summit.[2]

Article 2 establishes the pre-eminence of governments in decisions on the major aspects of the Summit. Only States had the right to vote in PrepCom meetings, and they therefore played a dual role at the WSIS by making contributions that expressed their interests and by exercising their power to vote on all of the key points up for debate. In this sense, the WSIS innovated little compared to previous UN events.

At the same time, the private sector had effective means of ensuring that its interests were represented at the WSIS. ICTs play an important role in economic development, and governments are always ready to lend an attentive ear to the private sector's requests. We have already mentioned the privileged role of the G8 Dot Force. The Organization for Economic Co-operation and Development (OECD) has bestowed the title of main promoter of information society-related research, development and education upon the private sector. The private sector's role in the actual implementation of ICTs is major, and is recognized by the States, which are seeking to attract as much investment as they can within their respective borders and to develop closer ties with the private sector:

> Almost all OECD countries have well-developed and clearly enunciated broad strategies and action plans for IT and an overarching policy approach to the information society. These usually cover technology development, technology diffusion, improving the IT environment and the global diffusion and distribution of ICTs. Policies to encourage broadband infrastructure investment and use are receiving more and more attention. The potential cost-effectiveness of public-private partnerships in promoting the development and use of ICTs is increasingly recognised.[3]

This desire to please private business is all the more evident because some States, especially those most affected by the digital divide, are in urgent need of direct investment. The market liberalization movements, both regionally and globally, are also in favour of representation of private sector interests at the WSIS.

Civil society was unquestionably the least favoured actor in the development of positions of strength at the WSIS. Civil society, which was invited to the Summit for its "pragmatism", its "experience in the field" and the legitimacy its presence would help confer upon the undertaking, faced many challenges that we will expand upon later. Suffice it to say here that civil society was confronted with problems of funding and internal structuring and with a lack of real means to exert pressure on the States. The Summit's rules of accreditation and participation prevented civil society from constituting the kind of force needed to effectively represent the interests it tried to defend within the official structures; consequently, a large portion of the energies its members invested at the Summit was devoted to improving its position of strength among the forces present.

NGOs at the United Nations

The inclusion and effective participation of non-State actors has become an important issue at UN conferences and events. The reforms the UN has undertaken with respect to the role and the place granted to the different components of civil society and the private sector are redefining the relationships that determine international governance. The motivations behind these reforms are pragmatic:

> One of the main reasons for civil society to participate in conferences and General Assembly special sessions is to contribute to the outcome of these meetings and take part in the implementation and follow up processes.[4]

Yet the participation of non-governmental organizations is not a recent phenomenon, as we can see from Article 71 of the Charter of the United Nations.[5]

> The Economic and Social Council may make suitable arrangements for consultation with non-governmental organizations which are concerned with matters within its competence. Such arrangements may be made with international organizations and, where appropriate, with national organizations after consultation with the Member of the United Nations concerned.

A number of States have been calling for many years for a fuller and more transparent inclusion of NGOs in international governance processes. Several major developments (globalization of markets, repeated requests for clear recognition of economic and social rights, increased worldwide awareness of poverty) have helped open the door to a more active presence on the part of civil society at the United Nations.

In 1993, UN member States decided to revise the rules dating back to 1968 that governed the inclusion of non-governmental organizations in the UN system. The revision was completed in 1996 with the adoption of Resolution ECOSOC 1996/31,[6] which remains in effect today.

This Resolution also establishes three levels of consultative status for NGOs. General consultative status is held by the major NGOs whose work covers a broad range of the concerns of the UN Economic and Social Council (ECOSOC). Special consultative status is reserved for NGOs which have a specific field of expertise in one of the sectors of the Council. The third category is made up of a group of NGOs whose expertise may occasionally be put to use in the work of the United Nations. All NGOs with consultative status at the ECOSOC are accredited to international meetings. Those that do not have consultative status must obtain accreditation from the Secretariat of the event in question.

The Secretary-General laid down the specific rules for NGO participation in Report A/53/170,[7] published in 1998 and complemented in 1999 by Report A/54/329.[8] The Millennium Declaration subsequently strengthened the institutional predisposition toward NGO participation at the UN by giving a new and expanded mandate to NGOs working in cooperation with the UN.

Kofi Annan's efforts to deepen and widen the partnership between civil society actors and the United Nations were further elaborated in the report entitled *Strengthening of the United Nations: An Agenda for Further Change* (A/57/387), in which the Secretary-General calls for the establishment of a panel of eminent persons *"to review the relationship between the United Nations and civil society and offer practical recommendations for improved modalities of interaction"*.[9]

The Panel of Eminent Persons on United Nations-Civil Society Relations was appointed in the beginning of 2003 under the chairmanship of the former president of Brazil, Fernando Henrique Cardoso. It delivered its report (A/58/817, also known as the Cardoso report) in June 2004. The Panel considers that *"civil society is now so vital to the United Nations that engaging with it well is a necessity"*.[10] The Cardoso report proposes four general principles under which the UN reforms must take place. The UN should thus:

- Become an outward-looking organization. The changing nature of multilateralism to mean multiple constituencies entails the United Nations giving more emphasis to convening and facilitating rather than "doing" and putting the issues, not the institution, at the centre.

- Embrace a plurality of constituencies. Many actors may be relevant to an issue, and new partnerships are needed to tackle global challenges.

- Connect the local with the global. The deliberative and operational spheres of the United Nations are separated by a wide gulf, which hampers both in all areas from development to security. A closer two-way connection between them is imperative so that local operational work truly helps to realize the global goals and that global deliberations are informed by local reality. Civil society is vital for both directions. Hence the country level should be the starting point for engagement in both the operational and deliberative processes.

- Help strengthen democracy for the twenty-first century. The United Nations should accept a more explicit role in strengthening global governance and tackling the democratic deficits it is prone to, emphasizing participatory democracy and deeper accountability of institutions to the global public.

This was the context in which the "multi-stakeholder" structure of the WSIS was elaborated. We can now look more closely at how the organizers of the Summit translated these considerations into concrete participative measures.

Integration of Non-governmental Actors at the WSIS

UN Secretary-General Kofi Annan described the climate that should characterize the relationship between the UN and civil society in a speech given in Sao Paulo in 1998:

> It stands to reason that the relationship between the United Nations and civil society has changed beyond all recognition. (...) If the global agenda is to be properly addressed, a true partnership between NGOs and the United Nations is not an option; it is a necessity. (...) Yet despite the growing manifestations of an evermore robust global civil society, the United Nations has been inadequately equipped to engage it and make it a true partner in our work. And so when I took up the position of Secretary-General and embarked on a quiet revolution to reform the United Nations, enhanced cooperation with NGOs formed a crucial theme in my proposals. This stemmed from recognition that our common work will be more successful if it is supported by all concerned actors of the international community.[11]

The Resolution on which the Summit is based explicitly acknowledges this recognition and inscribes it in a process of building cooperative networks among all of the significant actors involved in the information society in order to attain the objectives outlined in the UN's Millennium Declaration.[12] This goal stems from the complementarity of the respective competencies, powers and areas of action of the stakeholders. Resolution A/RES/56/183 thus reads, in part:

> Convinced of the need, at the highest political level, to marshal the global consensus and commitment required to promote the urgently needed access of all countries to information, knowledge and communication technologies for development so as to reap the full benefits of the information and communication technologies revolution, and to address the whole range of relevant issues related to the information society, through the development of a common vision and understanding of the information society and the adoption of a Declaration and Plan of Action for implementation by Governments, international institutions and all sectors of civil society,
>
> (...)
>
> Recognizing the need to harness synergies and to create cooperation among the various information and communication technologies initiatives, at the regional and global levels, currently being undertaken or planned to promote and foster the potential of information and communication technologies for development by other international organizations and civil society,
>
> (...)
>
> Encourages effective contributions from and the active participation of all relevant United Nations bodies, in particular the Information and Communication Technologies Task Force, and encourages other intergovernmental organizations, including in-

ternational and regional institutions, non-governmental organizations, civil society and the private sector to contribute to, and actively participate in, the intergovernmental preparatory process of the Summit and the Summit itself.[13]

The World Summit on the Information Society thus marked a turning point in the way actors are integrated into the official negotiation system of the United Nations. For the first time, a UN Summit has been given an organizational structure consisting of a number of components which bring together representatives of member States, the private sector, civil society and various UN agencies. Similarly, the clearly expressed desire of the Summit organizers to include these actors from the beginning of the preparatory process is something new at the United Nations. We must note, however, that the organizers' willingness ran counter to the States' way of seeing things, a point we will come back to in the second section of the document. At the same time, we must also note that another historic precedent was set at the WSIS: for the first time, private sector entities were allowed individual accreditation, in a departure from the standard accreditation procedure applied at previous Summits. In the past, only associations representing groups of private sector entities, such as the International Chamber of Commerce (ICC), have been accredited. The formula chosen for the WSIS is similar to the one used at the ITU, where individual private sector entities can sit alongside the States as associate members.

This raises important questions. ECOSOC rules prohibit the individual accreditation of members of the private sector on the principle that commercial enterprises do not fulfill a representative role and may have interests that are incompatible with the goals and interests of the United Nations. The private sector is represented by umbrella organizations such as the Coordinating Committee of Business Interlocutors (CCBI). Yet, the WSIS ignored the ECOSOC rule and has allowed the individual accreditation of private sector "entities". At PrepCom 1, commercial enterprises were considered on the same level as NGOs. This set a precedent in United Nations practice and changed the relationship established between the United Nations and civil society over the last fifty years. The ITU transferred to the WSIS its institutional practice of individual commercial enterprise accreditation. The private sector was now represented two-fold, sometimes individually and collectively. This new dynamic weakened civil society whose influence was diluted amid private sector interests. The lack of public debate over the inclusion of for-profit corporations revealed a deficient approach to transparency in the preparatory process. The accreditation of individual commercial enterprises also raised many questions about the legality of this practice within the UN framework.

The organizers of the WSIS felt that the role of civil society would be crucial in defining an equitable and development-centred information society. The role civil society would play within the Summit was thus politicized: it added further legitimacy to the WSIS by positioning specific issues based on people's needs. This was recognized by the governments themselves, as stated in a draft document: "*Civil society involvement is vital in the take-up and social acceptance of the Information Society.*"[14]

The Summit organizers also saw a fundamental role for the private sector, as the sector that would ensure the sustainable development of the infrastructures, content and applications of the information society. The private sector has been heavily involved in cooperative efforts with public institutions to promote the development of infrastructures and labour force training. By working to develop markets and integrating ICTs, the WSIS also saw the private sector as playing a political role.

The international agencies involved in the Summit were there primarily to provide expertise depending on the nature of their respective terms of reference. Their specializations and areas of jurisdiction made them key partners in the implementation of WSIS policies. Expertise was much sought after at the WSIS, which has a technical and complex subject to manage.

Non-governmental entities participated in the Summit within the framework established by the series of relevant resolutions adopted throughout the 1990s. While the WSIS organizers displayed a spirit of openness, the States which adopted the Rules of Procedure remained conservative. NGOs and the private sector were given consultative status at the WSIS. The preparatory process recognized *de facto* the diverse character of the Summit but also the predominant influence of States in determining the shape of the event, as decision-making power at the WSIS (active participation in the Subcommittees and voting rights) was reserved for the government delegations.

The WSIS therefore offered civil society and private sector entities the opportunity to:

- provide expertise in specific areas;
- raise concerns and issues and bring them to the Preparatory Committee discussions;
- advise the delegations on how to formulate and implement possible policies;
- develop new avenues of economic and social development
- lobby States;
- advocate for their respective interests internationally;
- influence the discussions leading to the drafting and adoption of the *Declaration of Principles* adopted in Geneva in 2003;

- contribute to the actual implementation of the *Plan of Action* adopted in Geneva in 2003.

The private sector seemed content with the role the WSIS assigned to it. Civil society, on the other hand, invoking the promises made in the Summit's enabling resolution, demanded a greater role in the deliberations from the outset. As we will see in Part II, these demands ultimately led to a significant political victory when an institutionalized role for civil society at the United Nations was recognized through the creation of a Civil Society Bureau at PrepCom 2.

Competing Visions

The interests of the various stakeholders were clearly articulated at the WSIS. But the Summit cannot be reduced to a forum where different positions were expressed; it was also an arena in which a political game was being played by the representatives of governments, the private sector and civil society. The game involved the building of alliances, in which each party attempted to make as many gains as possible depending on its respective areas of concern and interests. The objectives pursued by the stakeholders at the WSIS can be summarized as follows.

Governments

> All governments have a stake in the information society, whatever their level of national income or their infrastructure facilities. Governments are key for bringing the benefits of the information society to everyone through the development of national and global policies and frameworks to meet the challenges of the information society. In their pursuit of the public interest, governments can raise awareness, facilitate access to information for the public, and they also can lay the foundations for all citizens to benefit from Information and Communication Technologies in terms of improved quality of life, social services and economic growth.[15]

Government policies on the information society and on the Summit varied greatly depending on their respective national realities. The discussions in Subcommittee 2 on Content and Themes served to identify the general principles that guided the government positions and to narrow the range of themes they wanted to see considered at the WSIS.[16] The governments emphasized the themes that related to applications, network security and infrastructure development. Although some States showed a degree of openness to the demands of civil society, particularly with respect to communication rights and the social role of ICTs, the attention actually paid to these issues was limited.

It is therefore important to remain cautious when analyzing the elements included in the final documents. Governments tended to focus on some content at the expense of more fundamental but less "attractive" priorities. For example, much more attention would surely be paid to trade measures than to bridging the digital divide, and there would certainly be more interest in network security than concern over human rights. The subtle and seductive language of governmental principles veiled specific interests that not infrequently conflicted with the ethical statements in the documents. We will come back to this point in our critical analysis of the Summit at the end of this book.

Private Sector[17]

> The private sector plays an active role, in conjunction with governments and civil society, by offering an economically viable model to achieve the development objectives on the world agenda. The contribution of the private sector is instrumental in creating the material conditions for universal access to information and value-added ICT services. Its involvement in the Summit promotes economic growth and new partnerships, technology transfer, increases awareness of new technologies, and motivates the creation of local content development and skilled employment opportunities.[18]

Compared to other actors, the private sector maintained a relatively discreet presence at the WSIS. The private sector attended because it wanted to influence the economic policies of the States, to which it already has relatively easy access by all accounts. And there was no need for it to be overly assertive when some States, and the United States of America in particular, were all but mouthpieces for the private sector, which they see as playing a special role in their information society development strategies. The private sector does, however, have a clear view of the opportunities the WSIS has to offer.

The Coordinating Committee on Business Interlocutors (CCBI) and the International Chamber of Commerce (ICC) were the main groups representing the interests of the private sector at the WSIS. The International Chamber of Commerce chairs the CCBI, whose members are:[19]

- International Chamber of Commerce (ICC) – Chair
- Business Council for the United Nations (BCUN)
- Business and Industry Advisory Committee to the OECD (BIAC)
- Global Business Dialogue on Electronic Commerce (GBDE)
- Global Information Infrastructure Commission (GIIC)
- Money Matters Institute (MMI)
- United States Council on International Business (USCIB)

- World Economic Forum (WEF)
- World Information Technology and Services Alliance (WITSA)

According to the CCBI's spokesperson:

> This Summit and the preparatory process can provide a unique and essential mechanism to achieve global understanding of the policy, regulatory, and infrastructure issues that are critical to ensuring that the information society is accessible by all.[20]

In other words, the determining factors in ensuring universality of access to the information society are technical: the universality of the ICTs will be achieved through the development of infrastructures and a climate conducive to investment. This view is very strongly held in the private sector, and by some governments, led by the United States.

The vision, themes and contents dear to the private sector were very clearly articulated in a CCBI document submitted as a contribution to PrepCom 2.[21]

Point B of the section entitled *The private sector's key messages and expectations at the Summit* summarizes the essence of the corporate discourse at the WSIS: market liberalization and economic policies that favour competition, investment and the elimination of obstacles to trade. The conception put forward by the private sector suggests that social development will occur if specific economic measures are adopted. No mention is made of the digital divide or of social issues, although there are passages that argue that the business-friendly economic policies contribute to local development.

Civil Society

> Civil society is playing an active role in identifying the social and cultural consequences of current trends and in drawing attention to the need to introduce democratic accountability on the strategic options taken at all levels. Its diversity and, often, hands-on approach to issues, make civil society a key player in the renewed international partnership called for by the UN Secretary-General.[22]

As a heterogeneous group of entities whose allegiances and natures are extremely diverse, it is difficult for civil society to articulate a unified discourse, although particular concerns, themes and issues are common to many of its members. Civil society decided to get organized in a coherent manner at PrepCom 1. The Content and Themes Working Group which civil society formed at PrepCom 1 was intended to be the forum where the various positions came together in a well-articulated and coherent whole. The themes carried mainly by civil society revolved around sustainable development, democratic governance, literacy, education, research, human rights, common global knowledge, linguistic and cultural diversity, gender issues and "information security". Overall, civil society constantly advocated for an information society rooted in the concept of solidarity.

Civil society's basic premise was that the market alone cannot resolve the issues raised by the Summit. This being the case, a new direction is required, one that rests more on the application of criteria of ethics and social justice than on the development of markets. Although there are some complementary aspects to the two positions (the need for economic development, training and education on ICTs), differences of position remained, particularly regarding the role of the public sector in the information society, the concept of public good, the criteria for access (as a right rather than by market demand). As a result, the relationships at the WSIS between civil society, the private sector and the States seemed locked in a dynamic of convenience—each side needed the other, and if cooperate they must, they did so with mistrust.

PART TWO

Civil Society at the WSIS

The World Summit on the Information Society was above all a political event. The parties it brought together—civil society, government delegations and the private sector—did their best to maximize their respective gains, acting individually or through alliances. For civil society, this was an altogether new experience, in terms both of its organization and of the form of its participation. The following section presents a history and assessment of this experience.

• CHAPTER FOUR •
Participating in the WSIS

> In many areas, but especially those with broad ramifications across society, it is no longer conceivable that effective policy and programme and implementation can be achieved without active participation from civil society.
> —WSIS Executive Secretariat (2002)[1]

> Civil society is playing a key role in identifying the social and cultural consequences of current trends and in drawing attention to the need to introduce democratic accountability on the strategic options taken at all levels. Its diversity and hands-on approach to issues make civil society a key player in the renewed international partnership called for by the Secretary-General of the UN.
> —WSIS Brochure 1 (2002)[2]

Civil society mobilized as soon as announcement of a tripartite World Summit on the Information Society was made, and it lost no time positioning itself as a very involved actor in the preparatory process. The first scraps of information that became available sparked hopes of the dawning of a new era of negotiations at the United Nations, with the organizers talking about the possibilities of "*full participation of civil society and all actors*" in a new type of "multi-partner Summit".

The least one can say is that the initial responses were enthusiastic, albeit tempered by a hint of skepticism. Having been sidelined for decades, a rapidly expanding international civil society was at last going to be included in a global decision-making process. Questions as to the form its inclusion would take soon surfaced, since many organizations started out with very high expectations of the WSIS.

The various members of this new conglomeration referred to as global civil society began quite early on holding more or less formal meetings in order to organize their forces. Among other things, they sought to develop innovative relationships and strategies outside of the major UN institutions. New linkages were established. Many had already been in regular contact with each other for some time through gatherings such as the World Social Forum. Meetings were convened on the basis of particular criteria, mainly geographic, institutional (media groups, NGOs specialized in human rights, etc.), thematic (digital divide, communication rights...) or sectoral (youth, gender...). Quite a few of the organizations participated in more than one group.

The participation process showed how truly diverse civil society is. There were many differences of approach and work methods among the actors seeking to participate in the WSIS. While the focus, experiences and areas of concern varied considerably from one organization to the next, there were several dimensions on which they recognized each other and could agree. Generally speaking, civil society sought to take part—as a partner—by contributing its practical experience on the ground and a perspective on the issues of the information society that was both non-governmental and non-commercial. As the process unfolded, a common vision emerged: civil society at the WSIS took on the role of spokesperson for the voiceless and the have-nots, advocating on issues ignored in the dominant discourse and sharing its vision of an information society based on the principles of social justice, human development and human rights.

Many organizations and associations invested time and energy in Summit-related activities alongside the official events and on the edges of the dominant current in civil society. For example, they contributed greatly to the World Electronic Media Forum dedicated to discussing the future role of electronic media in the so-called information societies. The many conferences, round tables and discussion meetings gave a voice to a broad range of representatives of non-governmental and non-trade organizations. The Forum took place from December 9 to 12, 2003, in Geneva.[3]

The World Summit of Cities and Local Authorities on the Information Society was another significant opportunity for civil society to express its concerns and its demands. Local elected representatives, mayors, NGOs, community associations and private sector entities gathered on December 4 and 5, 2003, in Lyon in an atmosphere that was much more open than at the WSIS. The meeting produced the *Lyon Declaration*,[4] which was received by the Executive Secretariat as a contribution to the WSIS.

The adoption of the formula whereby the WSIS would be organized as a "daisy wheel Summit" opened the door to inputs from a wide range of peripheral activities in which civil society was very present.

The Catalyzing Role of the CRIS Campaign[5]

In December 2000, Mohammed Harbi, a senior executive of the ITU, told an international meeting of the Global Community Networking Conference in Barcelona that the participation of NGOs and other civil society associations would be crucial to the success of a world Summit that the ITU was organizing at the request of the United Nations. It is not clear in the name of what authority Harbi made this statement, but it certainly struck a responsive chord with the participants, who were unaware at the time that the WSIS was on the UN's agenda.[6]

In the months that followed, Voices 21, an informal association of media activists, practitioners and individuals formed in 1999 to educate the public on media and communication-related issues,[7] began thinking about how it could influence the outcomes of the WSIS. It was decided to revive the Platform for Democratization of Communication, a coalition of NGOs formed in London in November 1996 and including the World Association of Community Radio Broadcasters (AMARC), the Association for Progressive Communication (APC) and the World Association for Christian Communication (WACC).

On June 16, 2001, Reverend Carlos A. Valle, Secretary General of the WACC, wrote on behalf of the Platform to the WSIS interim coordinator Arthur Levin to ask him to organize a meeting in Geneva for the purpose of "*clarifying the opportunities for the involvement of civil society*" and generating ideas and possibilities regarding the process.[8]

The letter stated:

> Our WSIS NGO Working Group is beginning to formulate ideas on the participation of civil society, and also on the Themes under consideration at this important summit. The response to the WSIS from all members of the group is positive and enthusiastic, and indeed others with whom we are in communication also recognise immediately the importance of this event.

As no answer was forthcoming, the Platform decided to hold a meeting of its own in London in early November 2001 to move the process forward. At that meeting, the group decided to change its name to the Platform on Communication Rights and to launch a campaign for Communication Rights in the Information Society (CRIS), the purpose of which was "*to ensure that communication rights are central to the information society and to the upcoming World Summit on the Information Society (WSIS)*".[9] The declaration on the campaign's mission included the following points:

Our vision of the information society is grounded in the Right to Communicate, as a means to enhance human rights and to strengthen the social, economic and cultural lives of people and communities.

Crucial to this is that civil society organizations come together to help build an information society based on principles of transparency, diversity, participation and social and economic justice, and inspired by equitable gender, cultural and regional perspectives.

The World Summit on the Information Society offers an important forum to promote this objective. We aim to broaden the WSIS agenda and goals especially in relation to media and communication issues, and to encourage the participation of a wide spectrum of civil society groups in the process.

The connection that CRIS made between communication rights and the participation of civil society in a world Summit was not accidental. In fact, it was central to an even more fundamental connection between the issues and the processes that would mark the entire WSIS experience.

One of the key events of that period occurred a few days after the CRIS campaign was launched. The Platform joined forces with the German foundation Friedrich Ebert Stiftung (FES) which was working to bring a group of public broadcasters together to discuss the involvement of the media at the WSIS. On November 19-20, 2001, a meeting held by the Platform and FES in Geneva was attended by 35 people representing non-governmental organizations and media organizations in roughly equal proportions. The theme of the meeting was *Communication as a Human Right in the Information Society: Issues for the WSIS*.[10]

The meeting focused on the relevant issues and processes at the WSIS. On the second day, representatives of the ITU, UNESCO and the Civil Society Division of the WSIS Executive Secretariat took part and there was a wide-ranging discussion on the possibilities for civil society participation in the Summit. For the first time, participants were given a clear idea of what the WSIS would be about and how it would be structured.

The ITU and WSIS representatives stressed that there would be room for a significant involvement on the part of civil society and that this participation was seen as essential, despite the fact that a certain number of governments were less than enthusiastic about the idea. It was also clear that the WSIS agenda was still far from being finalized.

The CRIS campaign would go on to play a catalyzing role in helping to organize civil society at the first WSIS PrepCom in July 2002. PrepCom 1 was a time of great confusion and frustration for civil society's actors, as they could only take note that the doors of a so-called multi-stakeholder Summit were being closed one after the other in front of them. CRIS took on an unofficial

• Participating in the WSIS •   43

function of facilitating dialogue, mobilization and consensus building among the different organizations. In fact, the CRIS presence at PrepCom 1 was so strong that it attracted criticism from other actors who feared a subordination of civil society participation to CRIS's own agenda. After the event, CRIS stepped down from this informal role to refocus more on its own objectives. Several CRIS activists continued to play critical roles in the emerging civil society structures, however, as we shall see.

The Role of UNESCO

The formal participation of members of civil society in the activities directly associated with the WSIS began when UNESCO started consulting with NGOs in February 2002. At UNESCO's invitation, a number of civil society members met in February 2002 to identify the themes they wanted to see discussed at the WSIS. The organizations invited by UNESCO had specific expertise in the UN agency's areas of interest and represented various sectors of civil society. The discussions on the four themes for the consultations ("Infostructure" in Developing Countries, Cultural Diversity and Public Domain of Information, Freedom of Expression in the Information Society, Education In and For the Information Society) led to the drafting of inputs and recommendations on the principles that should guide civil society's participation at the WSIS. The four consultations brought together non-governmental actors with specific expertise in each of the four thematic areas. A fifth meeting was then held in April 2002 to broaden the discussion to civil society's participation in the WSIS. Some 100 representatives and 10 non-governmental organizations met to discuss the role and place of NGOs and civil society in the preparation and the holding of the Summit, and to propose points for inclusion in the *Declaration of Principles* and the *Plan of Action*.

It was during this series of meetings that civil society's hopes for participation at the WSIS reached their high point. Among other things, the text adopted at the fifth meeting in April 2002 stated that:

- A protocol for information dissemination and transparency in relation to the WSIS process should be explicitly agreed on and experimented with.

- Civil society should be represented on the Bureau of the Summit, as a means for participating in the ongoing preparatory process and to enhance its transparency.

- Criteria for NGO participation in the Summit should be clarified, in particular, to distinguish between lobbying organizations and other NGOs.

- A fund should be established to support effective and balanced civil society representation. States and donor agencies should be encouraged to contribute to this fund, which would be allocated, according to agreed upon criteria, by a competent agency, such as the United Nations Non-governmental Liaison Service (UN NGLS).

- Decentralization of the consultation and mobilization process could be reinforced by regional "animators" working within established NGO networks and properly resourced.

- Official recognition should be given to the consultations organized by civil society itself and mechanisms be made available to incorporate their outputs into the official process

- An ongoing mechanism for monitoring progress across all domains of the information society might be useful to identify ongoing and emerging concerns.

- The Summit should be accompanied by an NGO forum, in which civil society can discuss and organize input into the process and outcomes.[11]

A number of participants remained critical of UNESCO's participation procedures, which they saw as favouring institutionalized NGOs that are close to the United Nations and for the most part from Western countries. It is true that under the rules of procedure in effect at the United Nations, NGOs can be accredited individually for the international meetings to which they are invited. However, the extent of their participation is left to the discretion of the governments and can range from observer status to limited speaking rights. In the case of the WSIS, the call for a tripartite Summit sparked high expectations, especially among the groups and associations that were interested in the subject but unaccustomed to playing the UN game. Furthermore, for many of the more experienced actors, the time had come to insist on more decisive participation in the outcomes. Finally, the WSIS's definition of civil society suggested that participation would extend beyond the NGO sector.

The African Regional Conference

The official process leading up to the WSIS provided for a series of regional conferences and Preparatory Committee meetings to provide input for the Summit, the first of which was the Bamako regional conference in May 2002. After the consultations held by UNESCO, Bamako was the first opportunity for many civil society organizations to share experiences and start developing a common strategy that would enable them to have an impact on the preparatory process. It was soon evident that for logistical, tactical and strategic reasons, it was imperative that civil society become organized and structured, and that the capability of individual actors would be significantly strengthened if resources and expertise were pooled and coordinated strategies were developed. Bamako revealed both the opportunities and pitfalls of participating in the WSIS for civil society. Of all the official events organized as part of the WSIS preparatory process, the regional conference in Bamako was probably the one that sent the most positive signal to civil society by recognizing that it had a substantial role to play in the development and implementation of the information society and by strongly supporting its most cherished themes. The policies put forward by the regional meeting were the first major official input to the Summit.

PrepCom 1

The CRIS-FES seminar in Geneva, the UNESCO consultations and the Bamako regional conference were the first opportunities for many associations to meet, but civil society really started organizing for the WSIS at the first meeting of the Preparatory Committee, PrepCom 1, in Geneva on July 1-5, 2002.

NGOs and other civil society organizations accredited to PrepCom 1 (about 117 all told) arrived in Geneva without knowing exactly what kind of participation they would be granted. The Civil Society Division of the Executive Secretariat planned a symposium alongside the official deliberations where various civil society participants were invited to speak on different themes, but no framework was planned for discussions of a political nature dealing more specifically with the Summit as such. Faced with this situation, the proponents of a more active and more decisive involvement met on site to create such a space. These essentially spontaneous efforts gave birth to the Civil Society Plenary (CSP, also referred as "the Plenary"), a completely open forum which quickly became the place where civil society participants conferred and developed common strategies to maximize their influence on the preparatory process.

The meetings of the Plenary gave birth to a whole series of caucuses and working groups to deal with specific issues. The caucuses, based on common concerns (human rights, gender, sustainable development, etc.) or on common regional origin, became the primary producers of content agreed upon by the actors of civil society. They would forward their positions to the Civil Society Content and Themes Working Group (also formed at PrepCom 1) tasked with drafting coherent texts reflecting the full range of concerns of the members participating in the Plenary. The whole idea was to ensure a transparent bottom-up process.

The Content and Themes Working Group fulfilled a particularly important political function as a source of civil society input into the WSIS process. Its role was to bring together the contributions from working groups and caucuses in unified documents that expressed the demands of civil society with a more powerful voice and greater political legitimacy. It continued to function up to the end of the first phase of the Summit and, as we shall see, played a key role in preparing the independent Civil Society Declaration, *Shaping Information Societies for Human Needs*.[12]

The Plenary was established as a truly inclusive body in which all interested individuals and organizations were invited to participate actively according to their needs and interests:

> A key principle underlying the structures of civil society is that there must be multiple avenues and means for participation, and that all civil society entities can select the nature, level and extent of participation according to their needs and interests. Civil Society can constitute itself into a Plenary (CSP) at every official convening of the WSIS process, such as Regional Conferences, PrepComs, Intersessional meetings and Summits. The CSP is open to the participation of all civil society participants.[13]

Decisions at the Plenary (which continued to operate during the second phase of the Summit) were made by majority vote of the representatives present at the sessions. This limitation of participation to those in physical attendance set up a *de facto* discrimination against the more financially challenged or geographically distant organizations. The Plenary's representativeness problem was real, because at best it represented only those civil society entities which were accredited to the WSIS. The WSIS's prerequisite for accreditation was an additional constraint on civil society representation, although the number of accredited organizations rose significantly after PrepCom 1. The Plenary's openness to the full participation of accredited organizations rapidly lent it credibility as an interlocutor with the official bodies. The final CSP session at PrepCom 1 confirmed the Plenary's role as a decision-making body and functional tool of cooperative management. From then on, it was the ultimate source of common decisions on behalf of civil society at the WSIS. Its various constituent caucuses and working groups created at PrepCom 1 also continued to evolve and grow. Some 223 representatives of civil society attended PrepCom 1.

Still, many civil society organizations expressed great disappointment over the process and the official outcomes of PrepCom 1. As we mentioned earlier, the resistance of government delegates to broad participation on the part of civil society contrasted with the official discourse of the Summit's organizers.

Efforts to be inclusive were minimal, particularly as far as the political space and structures offered to civil society were concerned. Civil society's attempts to organize were largely determined by the difficulties it encountered in its attempts to participate in the official process.

Bruce Girard, a CRIS campaign member designated to express the CSP's views at the end of PrepCom 1, clearly articulated the participants' feelings:

> We came to the PrepCom because the issues related to the promised information society are fundamental to our concerns for social, economic, and human development, and because we believe that a vision of a people-centred information society can only be achieved with the full and active participation of civil society.
>
> We also came here because statements made by the UN Secretary-General, Kofi Annan, and Declarations and official documents issued by the United Nations, the ITU and the WSIS Secretariat repeatedly emphasised the need for the full participation of NGOs and civil society.
>
> (...)
>
> Major decisions faced by government had to do with how we would be able to participate in the official process, including such things as how and how often we would be able to address official sessions. However, the results of three days of meetings behind closed doors leave us with serious reservations. We had hoped for innovation. In their most optimistic interpretation, the agreements reached here represent a variation on established practices, but nothing in the way of positive innovation. Other interpretations see the decisions reached here as a major setback—eroding rights and responsibilities won by civil society in the UN system over the past fifty years.
>
> Encouraged by the various Declarations, announcements and official documents, we had hoped to be able to contribute to the process by participating in the organising bureau, joining in formal and informal agenda discussions, and having a voice in decisions concerning ongoing participation of civil society in the process. We hoped to be able to actively contribute new ideas to the partnership we were invited to join.
>
> What we got was disappointing.
>
> We will not be able to participate as observers in the Bureau. We can be excluded from participation in the agenda development. We have no guarantee of inclusion in significant aspects of the formal process.
>
> We are also disturbed by the possible precedent of accrediting individual firms to UN summits. The private sector has always been capably represented by its trade and industry associations, accredited by the UN as NGOs, but this summit is also proposing the formal accreditation of individual firms, responsible primarily to their shareholders or individual owners. A decision to include individual commercial actors in this manner in a UN summit, without the appropriate discussion and reference to established procedures, is unprecedented and we will be challenging it at the highest levels of the UN system.[14]

In summary, the outcomes of PrepCom 1 for civil society were mixed. Having been invited to the WSIS—thereby helping to legitimize the UN event—civil society thought it would be able to make a contribution. Its subtle exclusion from the debates and decision-making bodies would lead it to either support by default or dissociate itself from policies it was unable to influence. Paradoxically, one of the purposes of the WSIS was to capture civil society's perspectives on the issues surrounding the information society.

At the same time, a UN Summit is a broad forum in which civil society has an opportunity to globally consolidate and disseminate its principles and projects. This was of primary importance for many of the associations present at the WSIS. Important structures were established in Geneva in July 2002 as a result of civil society's efforts to maintain a high level of presence at the Summit. The events up to the end of PrepCom 1 created the conditions in which civil society organized its participation at the WSIS. They defined the framework in which civil society intervened as an actor at the Summit. This framework is the subject of the following chapters.

• CHAPTER FIVE •

The Organizational Structures of Civil Society

Before going any further, we shall review the structures officially established by the WSIS to encourage participation by civil society. There were basically two: the Civil Society Division (CSD), set up within the WSIS Executive Secretariat when preparations began, and the Civil Society Bureau (CSB, also known as the Civil Society and NGO Bureau), established at PrepCom 2 to facilitate liaison between the official organization of the WSIS and the autonomous structures of civil society that emerged from Prep-Com 1.

The Civil Society Division

The Civil Society Division was one of the four divisions that made up the Executive Secretariat. Its role was to facilitate and maximize the participation of civil society. In other words, it was the official transmission belt between civil society and the other WSIS stakeholders. The CSD was set up when work to organize the Summit began in late 2001, and was headed by Alain Clerc and Louise Lassonde, respectively director and coordinator of the Division.[1] As a coordinating structure supported by the Executive Secretariat, the Civil Society Division was intended to be a place where the many organizations that make up civil society could network. It also played a major instrumental and strategic role in the development of relations between civil society and the Summit.

The mandate of the Civil Society Division was to:

- brief all players on events and information pertinent to the Summit;
- provide civil society participants with the information and working materials necessary for their full inclusion in the preparatory process;
- inform other Summit participants of civil society's concerns;
- facilitate workshops and seminars on key issues affecting civil society;
- guide on-line discussion groups of civil society participants;
- work closely with the media to ensure that the issues of civil society will be heard;
- collaborate with other divisions of the Executive Secretariat;
- seek new perspectives on the subjects listed on the Summit's agenda.[2]

Although civil society participants greatly appreciated the dedication of the Division's staff, serious criticism was levelled at the way the Division discharged some of its duties. The Division was the official channel for the dissemination of information for civil society at the WSIS, but its job was problematic throughout the process. Some felt that information was transmitted ineffectively, there was too much reliance on electronic mail, participants were not always reached within reasonable timeframes and information was not always sent to the right people. To give but one example, civil society was informed less than two weeks in advance by the CSD of the informal meeting on content and themes held in September 2002. Many felt that civil society's communication with the media and with other actors did not receive the attention that it deserved.

From the outset, civil society's inputs were given little consideration in the official documentation of the WSIS. This was actually a reflection of the relative incapacity of the Civil Society Division to promote civil society's positions and content and a lack of appropriate resources. The chronic underfunding faced by NGOs at the WSIS was another of the difficulties the Division experienced in drawing attention to the needs of civil society within the Executive Secretariat. Ultimately, the Civil Society Division's importance lay in the role of secretariat it played for the civil society groups and its function as Secretariat of the Civil Society Bureau after the Bureau was created in February 2003.

The Civil Society Bureau

The creation of a Civil Society Bureau was above all the result of a long campaign by the Civil Society Coordinating Group (CSCG) to lobby the President of the preparatory process and the Executive Secretariat following PrepCom 1. Although the idea was controversial in all quarters, it was officially accepted by the Preparatory Committee at its second meeting (PrepCom 2) and ratified by the Civil Society Plenary on February 28, 2003.

The creation of the Civil Society Bureau has major implications for the inclusion of civil society in the political process at the global level. It is a first— never before have non-governmental actors been formally included in the deliberative structures of a global event of this sort. Taking its place alongside the Intergovernmental Bureau officially established at PrepCom 1, the Civil Society Bureau was the official interlocutor on the formal aspects of civil society's participation in the WSIS process. Since the role of the Bureaus is to facilitate the inclusion and representation of their constituent organizations in the Summit process, their function is procedural, in the sense that they perform duties relating to the administrative and organizational aspects of the preparatory process.

The Civil Society Bureau's primarily logistical role was to provide civil society with the structures it needed to effectively participate in the Summit. To this end, the Bureau sought to facilitate exchanges, better disseminate information and improve communications.

In the eyes of many observers, creation of the Civil Society Bureau represented a major political victory in the WSIS preparatory process. It was a breakthrough that could set a precedent for other international meetings. To quote the official announcement by the WSIS:

> The Civil Society Bureau is a decisive turning point in the history of the United Nations and of international negotiations. Indeed, it is the first time that civil society will have the means to effectively participate in the debate and will assume its responsibilities as a government interlocutor.[3]

This view was widely shared by civil society activists:

> The terrain the NGOs have gained is important. The only intergovernmental agency that had recognized the participation of independent sectors prior to this is the International Labour Organization (ILO), which has representatives of governments, businesses and trade unions sitting on its administrative council.[4]

Creation of the Civil Society Bureau also showed that the legitimacy and the role of civil society at the WSIS were eventually recognized. In other words, civil society won a victory of status. The Bureau is made up of civil society "families" clustered around common themes and each headed by a representative organization. The role of this organization was to ensure the most adequate representation of "family members" by virtue of its calibre and international character. Decisions were voted on by the representatives after prior consultation with their members.

The mandate of the Civil Society Bureau read as follows:

> Similarly to the Governmental Bureau, the functions of the Civil Society and NGO Bureau are organizational. Note that the Bureau is not responsible of the substance and cannot take position on substance on behalf of the families. Concretely, the CS & NGO Bureau makes decisions on organizational matters related to the various dimensions of the preparatory process and of the Summit, such as:
>
> 1. suggests how civil society entities should be represented at PrepCom and at the Summit and advises on all other questions related to organizational and practical arrangements pertaining to civil society participation in the process;
>
> 2. examines logistical needs and sets up operational guidelines for civil society events organization at PrepCom and at the Summit (time management plan, space, criteria for exhibits selection, etc.);
>
> 3. acts as the "connector" within civil society families to enhance mobilization and participation through information sharing and other initiatives;
>
> 4. sets the rules, criteria and mechanisms to allocate funds, identify organizations, networks and individuals who should benefit from fellowships; a selection committee will act as 'broker' for the different requests for funding;
>
> 5. examines and consults with the CSD /WSIS ES on the design of a concept for civil society involvement in the WSIS process;
>
> 6. examines the various information and documentation issued during the WSIS preparatory process, makes comments and suggestions and promotes maximum access to this information by civil society at large.
>
> 7. The CS & NGO Bureau meets between and during the PrepCom as appropriate. The Bureau convenes the Plenary meetings of civil society & NGO during PrepCom and at the Summit itself. To ensure efficiency, the executive members of the CS Bureau should be limited to 10 members, each representing one civil society family.[5]

The structure of the Bureau provided for the representation of 21 different families of which five represented the world's regions: Africa, Asia, Latin America and the Caribbean, Europe and North America, Middle East and Western Asia. Each family had one representative and grouped together various NGOs into a specific category based on the assumption that they shared a homogenous institutional culture, established or informal consultation mechanisms within the family group, and a common reference structure that could be well represented by an umbrella organization of international character. Communication mechanisms within each family which would allow adequate information dissemination and exchange between members of the group were deemed to exist.

Needless to say, the arbitrary division of civil society into "families" in this way was immediately subject to criticism.[6] The assignment of organizations, each with their own histories, priorities and ways of operating, to exclusive but vague categories such as "social movements" or "volunteers" raised questions as to the representativeness of the Bureau's membership. Bureau members did not have clear responsibilities and were not accountable to the organizations they represented. There was no specific mechanism to monitor the extent to which the coordinator of a family actually fulfilled his or her functions. In some cases, not-for-profit organizations found themselves grouped with commercial entities which took advantage of a lax application of the Rules of Procedure and the cover provided by the WSIS's broad definition of civil society. Issues of transparency and effectiveness thus made the Civil Society Bureau a source of tension as soon as it was established.

In fact, the creation of the Civil Society Bureau only strengthened the participants' feeling that independent and accountable structures were needed and increased their determination to establish and sustain such structures. The political decision-making body of civil society therefore remained the Plenary. The Bureau would play a logistical role but would be politically subordinated to the will of the Plenary. The Bureau would however play a critical role of liaison and communication with the Civil Society Division, the Executive Secretariat, the office of the President of the Preparatory Committee, and via them with the Intergovernmental Bureau. Established late in the day, it became one of the official operating constituent structures of the WSIS.

Chronological order	Origin	Functions
Civil Society Division Created in the fall of 2001	Executive Secretariat of the WSIS	• transmits information between civil society and the media, the Executive Secretariat and the other divisions of the Summit; • makes the concerns of civil society known; • secretariat of civil society.
Civil Society Plenary Created at PrepCom 1, July 2002	Organizations and groups of civil society present at the WSIS	• ensures exchanges between civil society organizations; • develops common strategies; • evaluates forms of participation and makes suggestions; • coordinates the activities and positions of civil society; • looks for means of funding; • ensures consistency within civil society (actions and positions); • works on targeted issues.
Content and Themes Working Group Created at PrepCom 1, July 2002	Organizations and groups of civil society present at the WSIS	• actively encourages the caucuses to draft and submit documents that clearly state their positions; • summarizes and incorporates the contributions of caucuses and working groups; • drafts balanced texts reflecting the positions of civil society organizations; • submits the common texts to the WSIS Secretariat according to the timetable.
Civil Society Bureau Created at PrepCom 2, February 2003	WSIS Preparatory Committee	• makes decisions on organizational matters relating to various aspects of the preparatory process and the Summit; • acts as "relay" between civil society families; • identifies logistical and material needs; • assesses the information and documents published during the Summit preparatory process, comments on them, suggests improvements and promotes access to the information by all of civil society.

Table 4. *Civil Society at the WSIS*

• The Organizational Structures of Civil Society • 55

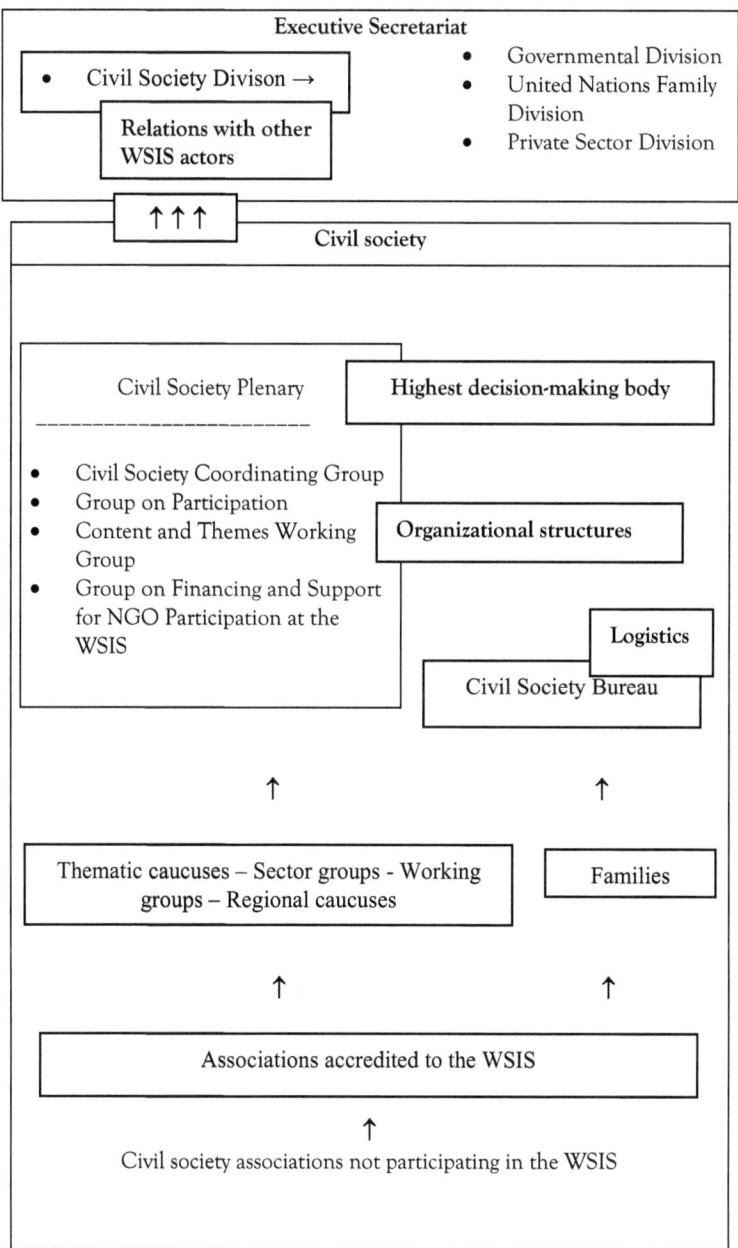

Figure 4. *Civil Society's Participation in the WSIS*

The Civil Society Plenary in All its Stages

As we have seen, the Civil Society Plenary emerged as a way to create a more equitable balance of power and to coordinate, if not unify, the different voices of civil society at the WSIS. Throughout the WSIS process, the Plenary and all of the sub-groups it created remained open and accepted new members on an on-going basis.

The Civil Society Plenary was formed during PrepCom 1 (July 2002). It did not see itself as exhaustive nor as the official voice of civil society, but acted as a political instrument open to all actors who wished to take part. The Plenary met only during the PrepComs and the WSIS itself, and constituted the collective decision-making body for civil society, although no association was really required to abide by its decisions. The Plenary sought to find and express common agreement on a range of organizational and thematic issues without claiming to represent the diversity of its members. Caucuses were continually formed by actors who wanted certain thematic or regional issues treated in a specific way. The Plenary also struck subcommittees and issue-specific working groups as necessary. These only represented their signatories, until specific positions were adopted by the Plenary as a whole.

At the end of PrepCom 1, members of the various groups, caucuses and subcommittees that formed the CS Plenary were asked to designate a delegate to a Civil Society Coordinating Group. The Civil Society Coordinating Group had two mandates; it coordinated the various Plenary activities and served as the liaison between the WSIS Executive Secretariat and civil society up until creation of the Civil Society Bureau at PrepCom 2. It thereby played an unofficial leadership role within civil society. The fact that members of the CSCG came from different groups and caucuses present in the Plenary lent them some legitimacy with the Secretariat and governmental delegations. Its editorial and synthesis work was highly political and therefore very important for all members participating in the Plenary.

The Content and Themes Working Group regrouped the themes, principles and plans of action that the various civil society groups and caucuses considered as priorities. The contributions were synthesized into documents that tried to coherently address as many demands in the inputs as possible. Much of its work took place on-line. Typically, a working document would be produced by a small drafting group and then be sent back to the various groups and caucuses for comments and criticisms. A final version would be submitted for approval before being sent on to the Executive Secretariat as an official input.

This kind of process had many advantages. First of all, it prevented government delegations from being overwhelmed by the weight of contributions from individual groups. It also avoided repetition within civil society (not to mention a waste of energy and resources), and provided all interested actors with a strong and unified platform. With the sheer volume of documentation, many civil society inputs would have been rarely consulted or become marginalized.

The adopted process allowed for production of a fairly complete and accessible overview of consensual civil society positions. Ultimately, of course, any civil society organization was free *not* to participate in the Plenary and its activities. But several hundred chose to take part.

Civil Society Caucuses and Working Groups	
Coordination and Strategy • Civil Society Plenary • Content and Themes Working Group • Civil Society Bureau	**Regional Caucuses** • Asia-Pacific • Africa • Latin America and Caribbean • Europe (EU, Candidate Countries, Switzerland) • North America • Western Asia and the Middle East • Arab Countries
Multi-Stakeholder Caucuses • Gender Caucus • Youth Caucus	**Thematic Caucuses and Working Groups** • Community Media Caucus • Finance Caucus • Health and ICT WG • Human Rights Caucus • Indigenous Peoples Caucus • Internet Governance Caucus • Media Caucus • NGO Gender Strategies WG • Patents, Copyright and Trademarks WG • Persons with Disabilities • Privacy and Security WG • Scientific Information WG • Trade Union Caucus • Telecentres WG • Values and Ethics WG • WG on Volunteering and New ITs

Table 5. *Civil Society Caucuses and Working Groups*[7]

On-line Communication

Communication technology itself was at the heart of a process to create a new on-line public space for civil society participants in the Summit. Independently developed networks were put in place as of PrepCom 1, making on-line communication critical in rallying the participation of civil society in the WSIS.

An electronic structure was created on the Internet to reflect the formal and physical organizational structure of civil society. There were basically three levels of on-line communication used by civil society at the WSIS: a Virtual Plenary listserv, the Content and Themes Working Group listserv and lists maintained by the various caucuses and working groups. Alongside physical presence at meetings, this became the main method of participation. As a palliative for chronic lack of funding and resources, on-line communication—an excellent tool for transcending geographic distances—became an essential organizational space for large international gatherings attended by civil society.

The SC Virtual Plenary

The establishment of a Virtual Plenary listserv coincided with the creation of the physical plenary established during PrepCom 1. The goal was to ensure a high level of transparency and the greatest possible access to debates and questions related to civil society in general. Everyone could contribute and the archives remain open and accessible.

The Civil Society Plenary physically met during PrepComs and at the Summit; it was not only the ultimate decision-making venue but also legitimated the participation of civil society in the WSIS. Rather than a space for decision-making, the Virtual Plenary was a complementary space where anyone (organizations officially accredited to the Summit as well as other interested parties) could express themselves without physically participating. The Virtual Plenary also acted as a link between the face-to-face meetings of civil society organizations.

The only requirement for participating in the Virtual Plenary was to agree with the following definition of civil society:

> Organizations—including movements, networks and other entities—which are autonomous from the State, are not intergovernmental or do not represent the private sector, and which in principle, are non-profit-making, act locally, nationally and internationally, in defence and promotion of social, economic and cultural interests and for mutual benefit.[8]

As a real public space dedicated to the WSIS, the Virtual Plenary developed a social link between actors and participants and also reinforced a sense of community and belonging. It integrated and recuperated the actors excluded from the Summit and renewed and developed relations between actors and organizations. Discussions took place around general issues related to WSIS participation, important Summit themes and official events connected to the Summit. In general, the Virtual Plenary generated openly shared information and acted as a large forum for planning and discussion as civil society actors attempted to develop a common Summit strategy.

CS Content and Themes List

The Content and Themes listserv was the main tool for generating debate on the development of common civil society positions. Caucuses were responsible for formulating texts yet their proposals could be modified or even discarded once they were debated on the list. There were discussions on the contribution process, policies to adopt and defend, the most effective text distribution strategies, and elements to be reviewed and modified.

Used as the main vehicle for discussing WSIS themes, the list relied heavily on feedback to formulate consensus-based civil society texts. The Content and Themes Working Group had the task of bringing together, synthesizing and organizing contributions from working groups and caucuses into coherent documents. Once texts were edited, they were submitted to contributors for comments, amendments and approval. This was followed by a new cycle of editing, all of this taking place on the e-list.

Archives were public to ensure transparency, which was fundamental to civil society planning on themes.

Caucuses and Working Groups[9]

As shown in Table 5, more than 25 caucuses and working groups were created by participants in the Civil Society Plenary. The difference between a working group and a caucus was never clearly established and had no real bearing on their respective activities. In general, a working group was considered to bring various actors together around a theme or specific issue such as copyright, scientific information, etc., whereas a caucus would gather a sub-section of civil society that shared a common identity, based on, for example, gender or youth. In some cases, however, the designation appeared almost arbitrary, as for example, in the case of the Human Rights Caucus, which focused on a theme that clearly touches on everyone.

The working groups and caucuses had to transcend physical distances and agree on positions to defend and agendas to prioritize, especially in between large meetings such as the PrepComs. They had to be able to make timely submissions and respond within sometimes very short time frames. This process took place on internal communication lists where general discussions were sometimes put on hold in order to discuss specific issues (and avoid clogging up the general lists). The internal lists were kept closed in order to keep strategic discussions proprietary and archives were only available to list members.

• CHAPTER SIX •

Civil Society Demands at the WSIS

Producing position and issue papers was one of the central activities of WSIS participants. This was certainly the case for civil society actors, who spent much of their time producing and disseminating common positions throughout the process leading up to the WSIS. One of civil society's key positions was its opposition to the term "information society", preferred by Summit organizers, in favour of the concept "information and communication societies". This choice of seemingly banal words set the battlefield for the Summit agenda.

This section demonstrates the collective efforts of civil society members in the development of common positions. Active members made significant efforts to develop collective materials—a testament to the effectiveness of civil society and its Summit structures.

The Battle for the Agenda

Establishing an acceptable power balance predetermines the ability of an actor to significantly influence the official agenda of a process such as the WSIS. Civil society quickly understood the relationship between participation and agenda setting and came up with a strategy to respond to these two kinds of issues.

The political game of large international meetings engages civil society into networking, sharing experiences and exchanging expertise. This makes it important to form alliances and create solidarity but also to develop common positions on shared issues. The strategies chosen at the WSIS were sometimes local, national or international, and were integrated formally and informally into the WSIS process.

For example, early on delegates expressed the need to lobby on a national level. They realized that the WSIS process was confusing for some; many government delegates only had an inkling about communication issues or approached the Summit based on the angle of the minister sent to the Summit. It was up to civil society to develop national networks, to encourage reflection and debate on the information society and to seize opportunities for influencing government policies. Many civil society actors actively worked on a national level in order to get "recruited" by their government delegations and to take advantage of a higher status in the official process.

The lobbying of government delegations at the WSIS happened in two complementary ways: formally in the context of official procedures, and informally by lobbying delegations more receptive to civil society demands. Governments like Canada, Costa Rica, Uruguay, Mali and France were immediately more open towards civil society and became potential allies. Although it may have been the main vehicle for action, official lobbying (the right to speak in plenary and the dissemination of positions and possible direction for actions) was not very efficient as there was little will to really include non-State actors. Despite this, the need to maintain a certain level of visibility and a strong official presence prompted civil society to invest greatly in official lobbying. This continued until November 2003 when the Civil Society Plenary broke from the official process.[1]

This kind of public relations work was not limited to the government sphere but also extended to other official spaces and to the Summit Executive Secretariat. The Civil Society Coordinating Group played a very important role in establishing good relations between civil society and the Executive Secretariat. Lobbying involved keeping pressure on the Executive Secretariat to effectively integrate the concerns of civil society and develop ways for civil society to participate in official events. These efforts produced positive results overall.

The Civil Society Plenary was never intended as a space to standardize participant discourses. Instead, it served as a space to create a coherent message and to develop solidarities relying on the strength of civil society to spread its points of view and address issues in distinct and innovative ways. In the end, it was able to build an important consensus around the declaration *Shaping Information Societies for Human Needs,* which was unanimously adopted on December 8, 2003, after some two months of collective editing facilitated by the Content and Themes Working Group.

As the Summit evolved, the real impact of civil society's many contributions on the official process was difficult to evaluate specifically, but it was clearly far from considerable. Certain efforts were made to analyze it.[2] Above all, the contributions produced at the WSIS were perceived to be an excellent opportunity for civil society members to get to know their respective organizations, their areas of work, their specific issues and points of view. Many organizations expressed the goal to develop their networks and converging interests at the WSIS.

In short, the structures put in place as of PrepCom 1 created a climate of exchange and fostered solidarity. In particular, the synthesis work done by the Content and Themes Working Group brought together and reformulated in an articulate manner relatively different issues on the spectrum of civil society concerns (for example, Internet governance and the role of communications in relation to human rights). Participants actively sought to generate general consensus or consensus within a community of interest as a means of political influence. As a result, the various working groups helped strengthen ties between civil society members, created and deepened reflexion on specific issues, and produced documents of greater significance.

The Main Obstacles

Civil society had to confront many difficulties in the process of formulating demands at the WSIS. These difficulties concerned both the political space created by the WSIS—including power relations, the openness of discursive spaces, the Rules of Procedure—and the internal capacity of civil society to deal with its diversity.

The Attitude of States

It was demonstrated previously that the UN framework does not lend itself easily to the inclusion of civil society. United Nations gatherings are primarily intergovernmental meetings; they can be attended by observers but are strictly framed by the concept that only nations can debate and legislate. From this perspective, civil society inputs are not considered much although they do carry some weight in the negotiation process. They are seen as actors who set directions for reflection and orientation that may or may not be addressed. The UN system operates on consensus: governments that are not very open to civil society inputs hold some power to limit their impact. Also, controversial positions from civil society or positions that risk affecting a powerful State have very little chance of being adopted in this political and diplomatic arena. Certain governments did not see the WSIS as an event tackling broader questions of communication, and preferred to concentrate on specifically targeted issues. The United States, for example, was only interested in three items on the WSIS agenda: network security, infrastructure development, and human capacity building. For them, it was out of the question to review the policies of international institutions such as the International Telecommunication Union (ITU), the World Trade Organization (WTO), and the Internet Corporation for Assigned Names and Numbers (ICANN). This kind of thinking was shared among a number of countries with strong ICT industries and whose visibility and power was much greater than civil society's. Clearly, this impacted the integration of civil society contributions to the WSIS.

Certain intellectuals, such as Alan Toner, see the WSIS as an arena where the global politics of information reflect the interests of dominant countries:

> As must now be clear, behind the WSIS' 'broad and general view' of 'knowledge dissemination, social interaction, economic and business practices, political engagement, media, education, health, leisure and entertainment' lies a very specific and ongoing set of strategies designed to use control of information, and information property, to advance Northern interests on the global scene.[3]

Overall, the political support most States offer to civil society is very limited in the UN setting. Certain States do not recognize the legitimacy of NGOs and seek to deny them effective participation in any official negotiations and discussions. In a venue like the WSIS, the vulnerability of developing countries and their urgent need for investment seems to encourage the adoption of conciliatory positions and leaves them receptive to the promises of technocratic discourse. In this context, it is not surprising that civil society had a difficult time moving its agenda forward.

Meryem Marzouki, the Human Rights Caucus coordinator, was mandated to speak on behalf of the Civil Society Plenary during the Intersessional Meeting in July 2003. She expressed a sentiment shared among various organizations in the Plenary:

> We have spoken about our suggestions with you, but we do not have the feeling we have been heard, or even listened to. Our legitimacy is not the same as yours, and we do not claim to be representative. Our legitimacy is anchored in our expertise, our field experience and our defense of a vision with public interest at its centre. We do not feel that this has been recognized or taken into account thus far.[4]

At the same time, some government delegations showed a real willingness to cooperate with private sector entities that enjoy enviable conditions for participation and are listened to by national representatives. Their individual accreditation to the Summit and their strong market-oriented vision, evident in the official documents of the first phase of the Summit, speaks to this problem.

Weaknesses of Civil Society

Although the diversity of civil society is considered a strength, it can also be a weakness on a political level. Developing various points of view and areas for reflection comes naturally to civil society, which fills an important social role through this process. However, in a political context this characteristic can weaken its position as it holds a relatively small concentration of power. Organization and political support are key elements in a political campaign but the WSIS participants found this difficult to achieve. Organizing through plenaries and structured working groups led to the unwanted consequences of diversity. There was a tendency towards concurrent discourses, repetition and the fragmentation of actors based on interests, which risked blurring civil society messages and reducing credibility. In effect, building a strong and coherent agenda demands an important investment in time and resources.

The lack of funding by the WSIS and the lack of means at civil society's disposal discouraged marginalized actors from developing strong positions and participating actively in their distribution. Civil society members who actively participated in the negotiations leading to the Summit were part of an elite group themselves; many organizations and NGOs based in the South were excluded from the Summit because there were almost no financial and organizational structures to enable their meaningful integration.

Content development efforts were often pushed aside during the first phase of the WSIS preparatory process as issues related to participation took priority. Nonetheless, civil society worked relentlessly to develop quality documents with the little financial and technical means at their disposal.

The low visibility of the WSIS in the media also kept civil society's positions from being taken into consideration. The mobilization of public opinion is one of NGOs' preferred means of pressuring national and international institutions. So, the indifference surrounding the organization of the WSIS meant that civil society could not rely on a political influence that might have been very helpful in promoting its agenda.

Raising Consensus

The real issues of the WSIS concerned its agenda. It goes without saying that the really substantial concerns were about themes and content. However, these often took a back seat to questions of procedure during the debates and to the power politics that characterized the Summit. Civil society owed it to itself to develop clear, well-substantiated, accessible and widespread positions in order to gain credibility as an actor that could be integrated into the process of negotiation and dialogue.

Indeed, the role of the Plenary and its various subcommittees, working groups and caucuses was for the most part, to encourage the production of thematic material and to ensure coherence in civil society's overall discourse.

PrepCom 1, July 2002, and its Follow-up

Civil society members seemed to recognize the gap in thematic material at PrepCom 1. After this meeting, additional efforts were made by the Content and Themes Working Group to fill that gap. Debate and discussion around the organization of civil society illustrated the different conceptions of, and variations in, the mandates of different members. So the real role of this Working Group became much larger than originally conceived; it became a key element in the production and diffusion of civil society positions.

The will to develop a more substantial agenda was stated many times in the period following PrepCom 1. A number of WSIS participants fine-tuned their approaches and concepts during the World Civil Society Forum (Geneva, July 2002). Notably, a WSIS working group was created touching on a number of themes: access to information in developing countries, governance and the information society, and the role of women in the information society.

As already indicated, this was a particularly tense period. There were no South-based NGOs present at the "informal meeting"[5] on content and themes that took place in Geneva, September 16-18, 2002. This meeting was held on short notice and for the most part took place behind closed doors. Nevertheless, the meeting sought to develop reflections on the information society, especially on principles and themes.

The European WSIS preparatory regional meeting in November 2002 also brought tensions to the surface between civil society and Summit organizers. The Civil Society Coordinating Group wanted a working session in order to develop the agenda whereas the organizers saw the regional meeting as more of a multi-stakeholder session, which would include the Coordinating Group in discussions. Given that its inputs had been given very little consideration up to that point, civil society had little interest in investing its resources on a formal meeting at a time when it was imperative that it develop its own consensus positions.

An on-line discussion forum moderated by UNESCO (December 2002-January 2003), also sought to bring clear positions into the WSIS preparatory process.

More than 300 NGO representatives registered for the UNESCO Forum.[6] The holding of the Forum itself created a certain amount of controversy.

There were reservations about the danger of reproducing the same inequalities that exist in face-to-face consultations on-line, namely: the exclusion of actors without access to the financial and technological means to participate effectively, the predominance of institutionalized NGOs, the repetition of the same discourses by the same people, and the marginalization of less powerful actors.

PrepCom 2, 17-28 February 2003

The PrepCom 2 contributions from the CSCG and the Civil Society Content and Themes Working Group were definitely more substantial than those produced previously. The official contribution[7] submitted at the second Preparatory Committee meeting targeted three areas (vision, principles, themes) and was a strong and structured input backed by the Plenary.

The Content and Themes Working Group also began to produce important documents synthesizing common civil society positions in the first three months of 2003. The first document of this nature was *Plan of Action: Civil Society's Priorities*,[8] a well-argued set of concrete proposals for actions and timelines for reaching specific objectives. The document was in fact a response to the *Draft Plan of Action* proposed by the intergovernmental Content and Themes Subcommittee, and mirrored the proposals found in that official document[9] with others reflecting the vision and concerns of civil society.

Civil society objectives were expressed in its most important content document produced during PrepCom 2, entitled *"Seven Musts": Priority Principles Proposed by Civil Society*.[10] This document was the outcome of a large on-line consultation among civil society members and synthesized a consensus among the actors who ratified the document:

Sustainable Development

An equitable information society needs to be based on sustainable economic and social development and gender justice. It cannot be achieved solely through market forces.

Democratic Governance

ICTs should facilitate democratic governance and foster participation by citizens. Transparent and accountable government structures at local, national and international levels should be established.

Literacy, Education, and Research

Only an informed and educated citizenry with access to the means and outputs of pluralistic research can participate in and contribute to Knowledge Societies. Access to tools and facilities that enable lifelong learning need to be created, extended and secured.

Human Rights

The existing human rights framework should be applied and integrated into the information society. ICTs should be used to promote awareness of, respect for and enforcement of universal human rights standards.

Global Knowledge Commons

Global knowledge commons and the public domain constitute resources that are cornerstones of a global public interest. They should be protected, expanded and promoted.

Cultural and Linguistic Diversity

Recognizing cultural development as a living and evolving process, linguistic diversity, cultural identity and local content need to be not only preserved but also actively fostered.

"Information Security"

"Information security" concerns should not infringe in any way on people's privacy and right to communicate freely, using information and communications technologies.

These "seven musts" grouped together civil society's major concerns as of PrepCom 2. Each point was divided into a series of sub-categories and encompassing issues such as Internet governance, communication rights, on-line governance, cultural development, gender, environment, the digital divide, etc. The guiding principles and fundamental points that underscore the "seven musts" are also found explicitly in the CSCG document entitled *Civil Society Statement to PrepCom 2 on Vision, Principles, Themes and Process for WSIS*.[11] This was without a doubt the most developed civil society document on content and themes during this period and was explicitly endorsed by at least 38 organizations. The visions, themes and principles outlined in the document represented the major demands of civil society members at the WSIS.

Intersessional Meeting, 15-18 July 2003

Governmental work did not advance sufficiently because of roadblocks that came up at the second Preparatory Committee meeting. As a response, Summit organizers convened an Intersessional Meeting in July 2003 in order to bridge PrepCom 2 and PrepCom 3 and give delegates more time to reach agreement.

The Content and Themes Working Group was also able to deliver a text synthesizing civil society's themes and priority issues (*Civil Society Priorities Document*) at the Intersessional Meeting in Paris (July 15-18, 2003).[12]

The document identified five common themes and four categories of issues as civil society priorities:

Common themes:

- Sustainable democratic development
- Human rights
- Global knowledge commons
- Literacy, education, and research
- Cultural and linguistic diversity
- Gender

Categories:

- "Information security" issues
- Access and infrastructure issues
- Global ICT governance issues
- Attention to other regional and international processes

The Civil Society Priorities Document illustrated the efforts of the Content and Themes Working Group to coordinate civil society inputs in a strategically effective way. The document was built on a consensus around fundamental points, established through a consultation process among different groups in the Civil Society Plenary. The five themes and four categories of issues in the *Civil Society Priorities Document* covered the spectrum of their concerns on the eve of the third WSIS Preparatory Committee meeting.

The issue of global ICT governance created serious debates in the heart of the Civil Society Plenary. Discussions focused mainly on what position to defend in relation to the large international organizations governing the Internet. These discussions were placed in a very specific context. The Internet Corporation for Assigned Names and Numbers (ICANN) and the International Telecommunication Union (ITU) were struggling on an international level over who would assume responsibility for Internet governance. The rivalry between the two institutions left civil society little room to maneuver.[13] Therefore, writing positions on ICT governance was sensitive and politically risky.

PrepCom 3, 15-26 September; 10-14 November; 5,6,9 December 2003

The third Preparatory Committee meeting marked a turning point for civil society. During the two weeks initially foreseen for the PrepCom, civil society participants realized that their most important positions had little chance of being included in the final Summit documents. Intergovernmental negotiations had become more and more difficult and there was even the possibility that the Summit would be canceled altogether due to the lack of consensus. Civil society representatives had limited access to the intergovernmental working groups created in an attempt to negotiate a consensus on contentious issues. On September 26, at the end of a marathon session lasting until 11 pm, the PrepCom suspended its work until mid-November.

During this period, civil society produced many documents responding to developments in governmental work. With the deadlines imposed by the WSIS approaching, the Content and Themes Working Group responded to the urgent need to participate more extensively in the official text contribution process with a flurry of texts. However, the limited impact of its contributions to the draft official texts prompted the Civil Society Plenary to break from the writing process, to withdraw its support from intergovernmental texts and to engage in a process of writing its own declaration. Before withdrawing, the Plenary did reach a consensus on the essential points that it considered non-negotiable. These were contained in a final civil society document that listed the priorities to be addressed in framing a definition of the information society. Civil society's last word on the subject could be broken down into ten themes: human rights, poverty reduction and the right to development, sustainable development, social justice, education and research, cultural and linguistic diversity, access and infrastructure, environment of good governance, global knowledge in the public domain, security and privacy.[14]

The World Summit on the Information Society, 10-12 December 2003

Governments continued to negotiate until the very eve of the Summit. Civil society followed these negotiations with a certain distance and worked on its own declaration as well on the "side events" where many of its most active members participated during the week of the Summit.

During this time, civil society finished its masterpiece, the declaration *Shaping Information Societies for Human Needs*. Unanimously adopted at the Civil Society Plenary on December 8, 2003, this was the most important text produced by civil society during the first phase of the WSIS. The document reflects a significant consensus on the fundamental elements defended by civil society and serves as a strong political input lending credibility, visibility and consideration to civil society. The main principles expressed are divided into four sections reflecting the collective views of the Plenary on the definition of a more human information society. A synthesis document was also produced in order to make the text more accessible to the media.

The collaborative process in which the declaration was produced was a success in itself. Under the coordination of the Content and Themes Working Group and rounding up the contributions of all of the caucuses and working groups who participated in the WSIS preparation, the declaration was essentially produced on-line during a period of around two months. The success of the endeavour demonstrated the motivation and maturity of civil society; it was able to reach consensus on a vast array of themes in a very short period of time, with no resources and in restricted conditions. The experience contrasted remarkably with the stalled intergovernmental process, which took place during the same time period.

Civil Society's Content Development Efforts

Legitimacy is at the foundation of all political exercises and the participation of civil society in the WSIS was no exception. Civil society was supposed to represent the forgotten voices, to present issues ignored by official discourse and submit different and innovative ways of doing things in order to defend the interests of marginalized populations. Yet the legitimacy of civil society was criticized early on in the preparatory process. This is a recurring problem in large international meetings and within progressive organizations in general. Most civil society spokespersons do not come from places where the needs and the issues are located. On the contrary, they enjoy the advantages of being highly interconnected and integrated in global networks.

The NGOs present at the WSIS were generally made up of professionals reflecting geographic, economic, demographic and gender imbalances. The least represented social groups at the Summit were precisely those who are cruelly suffering the inequalities that the WSIS sought to address. African NGOs were largely absent from the official process with a scant participation at best. This is not a phenomenon specific to the WSIS but is a recurring problem at large international events. It should be said, however, that civil society worked very hard from the beginning of the WSIS process to ensure the most equitable and balanced representation possible among its participants. Grants were offered to organizations of the South, and positions were allocated in regard to regional and gender criteria.

The absence of adequate WSIS financing for organizations based in the South significantly reduced their participation in the main events and in decision-making spaces. The themes of poverty, sustainable development and North-South solidarity were raised mostly by intellectuals and professionals of Western organizations. The South had little opportunity to make the voice of civil society heard in a significant way.[15]

Nevertheless, civil society structured actors into coherent groups capable of producing specific documents full of important concerns upstream or downstream of what was happening in the official deliberations. Certain groups such as the Youth Caucus, the Human Rights Caucus and the Gender Caucus were particularly active and effectively contributed to developing civil society inputs. Cooperation with the Civil Society Division also improved over time.

Overall, civil society succeeded in seizing opportunities to produce quality documents, although not without criticism. Two questions continually marked its debates: representation and legitimacy.

The problem of representation remained an unresolved issue from the creation of the Plenary. In whose name did participating organizations speak? In their own names or in the names of the people affected by their organizations? Or, better yet, did they consider themselves to embody a global civil society? Whatever the case, the value of any kind of representation was subject to debate. It remained unclear, to civil society members and to other Summit actors, in whose name civil society positions were formulated. This made it more difficult to understand the nature of the interests being defended.

Many questioned the legitimacy of civil society demands at the WSIS. Governments benefit from a certain political legitimacy and the private sector from economic legitimacy—though both of these are debatable—, but what civil society stands for remains ambiguous. It is recognized for having the necessary expertise and experience on the ground for implementing policies but not everyone agrees that it has the legitimacy to demand participation in their development.

Competition between civil society and government representatives emerged as of PrepCom 2 and continued through to the end of WSIS Phase 1. The question of legitimacy in representing the public interest was at the heart of this competition. Governments reflected their vision through the *Declaration of Principles* and the official Summit *Plan of Action*, and civil society through the contributions synthesized in its autonomous declaration. Governments and civil society actors competed to legitimize their respective conceptions of the public interest.

In fact, the problems linked to representation and legitimacy of civil society at the WSIS fit into a larger context of the multiplying number of transnational NGOs and the reshaping of international governance. Public space is being redefined in a number of countries accompanied by the emergence of new forms of political participation. New actors, different from traditional public and private players, are trying to legitimately address their discourses in a growing number of forums and to participate in the management of public affairs. It goes without saying that their effective inclusion in the processes of power distribution raises many questions.

Major Stages in Content Development

Civil society was not at the forefront of content development during the first year of the official WSIS preparatory process and then played a reactive role in regards to government inputs. In fact, civil society spent most of its energy on issues related to its own participation in the WSIS process. This situation changed as of PrepCom 2.

The Content and Themes Working Group played an important political role in bringing the many contributions from different groups and caucuses together into coherent and well-documented texts. PrepCom 2 was the space where civil society really became visible and began circulating major input in terms of content. The advances gained in terms of participation in the meeting itself as well as the improved level of organization and coordination allowed civil society to develop common positions more effectively.

The problems related to the definition of civil society weakened these contributions. Adama Samassékou, the Preparatory Committee President, recognized the problem surrounding the lack of clarity on the definition of civil society:

> Civil society is a very open concept that encompasses many different realities. The UN has a tradition of codifying NGOs but nothing similar for civil society. We should therefore study this reality and find satisfactory and constructive modes of operation together. In other words, it is necessary to come up with a typology of civil society actors that fits into the context of the Summit. It is crucial that all civil society members can participate in the Summit on the basis of their own modes of representation.[16]

The consultation and consensus-seeking process within civil society took the first of two important turns during the Intersessional Meeting in July 2003. This was a period when civil society had to face a number of fundamental setbacks in terms of the content of the official texts in progress.

In particular, civil society deplored the setbacks on communication rights, gender issues, democratic Internet governance, the public domain, cultural diversity and open source software. Governments were mainly concerned with questions of copyright and intellectual property rights sidestepping critical issues for civil society. The NGO Gender Strategies Working Group responded strongly to the suppression of paragraphs related to gender issues in the *Draft Declaration of Principles*. The Human Rights Caucus also moved into intense lobbying in order to keep human rights and fundamental freedoms at the core of the working documents. Within this atmosphere of tension, the Content and Themes Working Group was able to produce the consensus-based *Civil Society Priorities Document*. The obstacles faced by civil society actors encouraged them to seek consensus in order to hold more political weight with other WSIS parties. This will for consensus continued until the Summit itself.

The second turn was officially taken on September 26, 2003, when civil society gave an ultimatum to government delegations, in a press release stating: *"if governments continue to exclude our principles, we will not lend legitimacy to the final official WSIS documents"*.[18] This position led to the decision to prepare an independent Civil Society Declaration.

Important Thematic Documents Produced Collectively by Civil Society in the Framework of the WSIS I Preparatory Phase[17]	
PrepCom 2, 17–28 February 2003	*"Seven Musts"*: *Priority Principles Proposed by Civil Society*
	Plan of Action: Civil Society's Priorities
	Contribution on Common Vision and Key Principles for the Declaration
	Contribution on List of Issues for the Declaration and the Plan of Action
	Civil Society Statement to PrepCom 2 on Vision, Principles, Themes and Process for PrepCom 2
Intersessional Meeting, 15–18 July 2003	*Civil Society Priorities Document*
PrepCom 3, 15–26 September 2003 10–14 November 2003 5,6,9 December 2003	*Civil Society Essential Benchmarks for WSIS*
WSIS, 10–12 December 2003	*Civil Society Declaration: Shaping Information Societies for Human Needs*
	WSIS Civil Society Declaration Highlights
	Civil Society Essential Benchmarks (refined, 12 December 2003)

Table 6. *Documents Produced by Civil Society*

• Civil Society Demands at the WSIS •   75

Obstacles to the Content Development Process

The search for consensus implied a large amount of discussion and involvement by participating actors. This was challenging because civil society is made up of heterogeneous organizations, which sometimes have contrary sets of experiences and opinions. The WSIS framework was therefore a space where different paradigms and ideologies were continually confronted. Developing common positions necessarily involved facing opposite approaches and sometimes triggered important conflicts.

During the Intersessional Meeting in July 2003, to take one significant example, some members of the Media Caucus publicly attacked the CRIS campaign and those who supported the concept of the right to communicate. This was symptomatic of a more profound division that we will return to later:

> The polarization of opinions, which seems to date back to the UNESCO debates around the New World Information and Communication Order (NWICO) during 1970-1980, created divisions within civil society and divergences between the Content and Themes Working Group and the Media Caucus.[19]

The process of document editing itself undermined the legitimacy of some of the texts. Certain civil society members criticized what they considered to be an anti-democratic and non-transparent process. Contributions had to frequently be submitted by electronic mail within a very short time span, did not always receive enough feedback, and were subject to modifications considered by some to be arbitrary.

For some, the legitimacy of the Civil Society Declaration itself was questionable:

> This Declaration, was "adopted unanimously by the WSIS Civil Society Plenary on December 8, 2003." This boast is relative when considering that there were no more than 60 [sic] members of "WSIS civil society" present to approve this document by acclamation, whereas organizers accredited more than 1300 [sic] associations to this Summit.[20]

Civil society's organizing process around content was admittedly far from perfect. Decision-making remained problematic as flaws with regard to transparency, accountability and legitimacy dogged the Civil Society Plenary. Yet the accomplishments of civil society around content were numerous. Considering the technical, logistical and political difficulties faced in the production of consensual political documents, the lack of financing and appropriate resources and the attitude of many governments, the ability to remain a credible and coherent interlocutor was in itself a vital victory for civil society.

• CHAPTER SEVEN •

Themes Raised by Civil Society

Certain themes became particularly important for civil society groups during the preparatory process leading up to the Summit. A number of concepts were developed gradually and became core demands and priorities. The following is a snapshot of the most important civil society themes at the WSIS.

Financing the Information Society

As a major WSIS issue, the digital divide was a highly controversial subject throughout the preparatory process. All the players present at the WSIS agreed on the need to redress the gap but there was no consensus on how to make it happen.[1] The statistics are now common knowledge nevertheless: at the beginning of the 21st century, 91% of Internet connectivity was concentrated in the hands of 19% of the global population; there are more telephone lines in Manhattan than in sub-Saharan Africa; less than 1% of Africans have Internet access. In this respect, civil society and Heads of State alike share the same opinion: access to knowledge through new technologies is one of the keys to sustainable development.

Eliminating the digital divide requires monetary financing and a serious commitment by rich countries towards the development of communication networks in countries of the Global South. States therefore hesitate on what path to follow in order to achieve this goal. However, the proposal of Senegalese President Abdoulaye Wade to create a Digital Solidarity Fund (DSF)[2] received a positive response from civil society members who officially supported the idea.

The position of civil society on the financing of digital solidarity was distributed at PrepCom 3 in a document entitled *Civil Society Statement on Information and Communication Solidarity Funding Mechanisms*.[3] It stated:

For both new and existing mechanisms, we believe that serious consideration should be given to the following concerns:

Mechanisms for the distribution and implementation of such funds must be fully transparent and accountable, and ensure that the funds will reach the people who need them. The cost of administering the fund needs to be kept to a minimum.

- Fair rules for distributing international telecom charges based on solidarity must be re-established.
- Funding must not lead to further indebtedness and dependency to unequal trade relations, nor to privatization and deregulation. Funding must not lead to further reinforcing private and public monopolies.
- Contributions to any newly established fund must reflect a multi-stakeholder participation, providing contributions from governments and the private and commercial sector as well as from citizens.
- Funds must be managed and administered by people of the South.
- Gender parity must exist in all such mechanisms and all spheres and in all levels of decision-making and implementation.
- Funding mechanisms should not be established under the simplistic notion that digital divides can be redressed mainly by addressing issues of infrastructure or connectivity. They must be accompanied by funding for education, applications, content, and dissemination.
- In order to ensure that the funds actually reach the most marginalized, mechanisms need to be set up to ensure distribution of these funds take into account intersectionality of race, class, gender, ethnicity and other lines of discrimination.
- Must address traditional and community media, not only the Internet. It should take into full account low-end but appropriate technology, including indigenous knowledges.
- Must also be used to promote cultural and linguistic diversity.
- Must support technological sustainability, including the use and production of free software and the fostering of user-centered technology development practices.

The DSF meanwhile proposes a series of voluntary actions and initiatives to reduce the digital divide:[1]

This commitment is aimed at people, companies, associations, institutions and national and international organisations and the public authorities. It can assume two forms:

- public authorities (local and national) commit to include in their public bids for ICT (hardware, software and services), a clause for digital solidarity which stipulates that the company who wins the contract must make a contribution of at least one percent of the amount of the transaction to the Digital Solidarity Fund (in accordance with the "Geneva principle") ;

- as an alternative, public authorities (local and national) commit to donate an amount of at least one percent of their budgets earmarked for the purchase of ICT materials and services directly to the Digital Solidarity Fund.

A commitment similar to that undertaken by public authorities can be subscribed by private companies, citizens and any other institution interested to contribute.

In addition, public authorities, private companies, citizens and other interested institutions can decide to make :

- a commitment "in kind", through the intermediary of the Fund, donating equipments, software or free training ;

- a voluntary financial contribution paid directly to the Digital Solidarity Fund.

The Digital Solidarity Fund project is discussed in the Civil Society Declaration at point 2.4.4, *Financing and Infrastructure*.

Human Rights

The WSIS paradigm takes as a premise that the world is undergoing the impact of a third industrial revolution—the digital revolution—and that all social, cultural, economic and political relationships are affected. The title of the Summit itself indicates this: the event is about the "information society", a term considered nebulous and imprecise by numerous actors. The *Draft Declaration of Principles* of June 5, 2003 defined the information society as follows:

> The information society is a new and higher form of social organization, where highly-developed ICT networks, equitable and ubiquitous access to information, appropriate content in accessible formats and effective communication must enable all the people to achieve their full potential, promote sustainable economic and social development, improve quality of life and alleviate poverty and hunger.[4]

A strong ideological bias is present in this definition, at a time when 2 billion human beings still do not have access to electricity and where the great majority of the world's population has never made a phone call.

Civil society organizations were very involved on the theme of human rights, making numerous contributions and placing many resources at the disposal of partners. Notably, an on-line portal was created—*A Human Rights Portal to the World Summit on the Information Society*[5]—containing the principal themes related to the subject and a series of relevant resources. The Civil Society Human Rights Caucus was an active and substantial contributor to the WSIS.

An important human rights input at the WSIS was also produced at the International Symposium on the Information Society, Human Dignity and Human Rights, which took place in Geneva on November 3-4, 2003. The meeting resulted in a *Statement on Human Rights, Human Dignity and the Information Society*.[6] The text supports a series of rights considered particularly important in an information society: freedom of expression and information, non-discrimination, equality between men and women, the right to privacy, the right to equitable justice, protection of moral and material rights on intellectual creations, the right to participate in cultural life, minority rights, the right to education, the right to an acceptable quality of life, to health, to suitable nutrition, and to decent housing.

However, civil society's most significant contribution to human rights at the WSIS is contained in its Declaration *Shaping Information Societies for Human Needs*. Here, civil society projects an information society founded on the international human rights regime, reaffirming and reinforcing the universality and indivisibility of the rights proclaimed in the great international texts of the United Nations. This vision is expressed in the Declaration's point 2.2 *Centrality of Human Rights*:

> An information and communication society should be based on human rights and human dignity. With the Charter of the United Nations and the Universal Declaration of Human Rights as its foundation, it must embody the universality, indivisibility, interrelation and interdependence of all human rights—civil, political, economic, social and cultural—including the right to development and linguistic rights. This implies the full integration, concrete application and enforcement of all rights and the recognition of their centrality to democracy and sustainable development. Information and communication societies must be inclusive, so that all people, without distinction of any kind, can achieve their full potential. The principles of non-discrimination and diversity must be mainstreamed in all ICT regulation, policies, and programmes.[7]

The Civil Society Declaration was written out of discontent with the official documents. One of its strategies was to demonstrate the failings of the intergovernmental texts:

> Whereas the governments hesitated to reaffirm long agreed-upon human rights standards in their Declaration, the Civil Society Declaration develops ideas and strategies on how to realize, fulfil and bring forward the development and human rights of all people from a social justice focus.[8]

The following document illustrates this with reference to the climate of tension prevalent at the Intersessional Meeting in July 2003. Here, the Human Rights Caucus made an eloquent effort to keep solid references to international human rights standards in the official Summit documents.

> Exactly 10 years ago, in Vienna at the World Conference on Human Rights, over 170 governments reaffirmed their commitment to human rights. As governments, you collectively stated that human rights were universal, indivisible, interrelated and interdependent. You agreed that their protection was the first responsibility of governments.

Today, as we debate the challenges of the Information and Communication Society, let us not forget what has already been agreed upon.

Human rights are not a sectoral issue, relevant to only certain stakeholders. Human rights are one of the essential purposes of the United Nations, according to its Charter. The advent of information and communication technologies offers both opportunities and threats for those rights. If this Summit fails to reaffirm the centrality of human rights to its deliberations, we will have not only missed an opportunity, we will have sacrificed the civil, political, economic, social and cultural rights that we all hold dear.

The Human Rights Caucus welcomes the references to human rights in the draft *Declaration of Principles*. However, human rights should figure prominently throughout both the *Declaration of Principles* and the *Plan of Action* and WSIS should concentrate its efforts on devising concrete strategies to see that the rights recognized in international law are effectively implemented.

We support the reference to Article 19 of the Universal Declaration of Human Rights in Paragraph 10 of the *Draft Declaration of Principles*. This article must not only be affirmed, but also enforced. As we sit here in Paris, countless individuals in various parts of the world are detained simply for exercising, often using ICTs, their basic democratic rights to freedom of expression and freedom of association.

Privacy, a human right enshrined in international law, encounters specific challenges with the introduction of ICTs. Its protection will require strong language in Paragraph 52 of the *Declaration of Principles* and Chapters 5 and 6 of the *Plan of Action*. Privacy is not an ethical or moral issue; it is a fundamental human right.

In the name of a war against terrorism and with the pretension of increasing our security, human rights are being violated: right to a fair trial, presumption of innocence, equality before the law, freedom of assembly and association, freedom of movement and freedom from all forms of discrimination.

We don't need a culture of security, we need to ensure the security of cultures. That means that everyone, without discrimination, must be able to freely exercise their cultural rights and to use their own languages. ICTs will facilitate the realization of the right to education and knowledge only if fair and equitable access is within reach. That is the essence of a right to communicate.

"Everyone is entitled to a social and international order in which the rights and freedoms set forth in this Declaration can be fully realized". That is the promise of Article 28 of the Universal Declaration of Human Rights. The World Summit on the Information Society must not betray it.[9]

The Human Rights Caucus worked intensively to avoid a major setback in human rights at the WSIS in respect to other global UN gatherings. In the end, it helped "avoid a disaster" and also succeeded in including some important elements in the official documents.

The Right to Communicate

> The time will come when the Universal Declaration of Human Rights will have to encompass a more extensive right than man's right to information... This is the right of men to communicate.
>
> —Jean D'Arcy (1969)[10]

The right to communicate is a controversial concept that polarizes positions on the international information and communication system and goes beyond the Universal Declaration on Human Rights and the major texts associated with it.

> This right is perceived by its protagonists as more fundamental than the information rights presently accorded by international law. The essence of this right would be based on the observation that communication is a fundamental social process, a basic human need and the foundation of all social organization. The right to communicate should constitute the core of any democratic system.[11]

According to proponents of a right to communicate, international law does not yet consider the fundamental characteristic of exchange between individuals. Article 19 of the Universal Declaration of Human Rights presupposes a unidirectional transmission model that needs to be surpassed by integrating the profoundly interactive, feedback-oriented and participatory character of communication in order to affirm its eminently social essence. Effective access to individual or collective communication processes would therefore be a fundamental part of a right to communicate.

The idea of a right to communicate was put on the backburner of the international relations agenda following the heated NWICO debates at UNESCO in the 1970s and 1980s. The idea was never fully shelved, however. Pekka Tarjanne, then Secretary General of the International Telecommunication Union, notably suggested in 1992 that the Universal Declaration of Human Rights be amended to include the right to communicate. Kofi Annan, Secretary General of the United Nations, has also affirmed his support of the concept stating "*that millions of people in the poorest countries are still excluded from the 'right to communicate', increasingly seen as a fundamental human right*".[12] Responsible for organizing the WSIS, the ITU itself recognized the mandate to ensure the effective implementation of the right to communicate. This concern is mentioned in some of the official Summit literature.[13]

Advocacy for the right to communicate also fits into a critical perspective of negative media trends: monopolization and concentration of ownership, excessive commercialization, disinformation and content manipulation, information war and lack of pluralism. It is about reappropriating communication as a fundamental social process at the base of human society and as a tool for social and economic development.

Certain objections to the conception underlying this emerging right are very important to understanding the issues it raises. Numerous voices have expressed concern that opening up the Universal Declaration on Human Rights to amendment would be a very dangerous process. The climate following the Second World War was particularly ripe for the adoption of a progressive declaration; it is a climate that has disappeared today. Re-opening the Declaration could entail important setbacks. The text itself constitutes a coherent and structured whole yet modifications could risk creating incoherence or reinterpretation of content.

Also, the extremely wide interpretation that can be given to the right to communicate may make it difficult to apply consistently. Many organizations believe it is more important that a right to communicate by practiced on the ground rather than be abstractly formalized in international texts.

The CRIS campaign took the lead as the major promoter of the right to communicate at the WSIS, challenging the dominant international communication model which seeks to limit government intervention in communications to a minimum. CRIS spokespersons affirmed that it was never the intention of the promoters of the right to communicate to focus exclusively on its integration in international law. According to them, the issue is not about exposing gaps in the Universal Declaration on Human Rights but developing a larger strategy to encourage awareness on the subject.

The right to communicate thus supposes the intervention of a certain kind of regulation, or at least the active participation of States through the provision of a legal and regulatory framework in order to implement the right on the ground. This was a key element in the divergence that surfaced at the WSIS between groups dedicated to the freedom of expression and those who supported the right to communicate.

The concurrent visions of CRIS and ARTICLE 19—an NGO that advocates for freedom of expression—created tensions within civil society. In winter 2003, a text by Cees Hamelink, professor of international communication at the University of Amsterdam, circulated among CRIS members with the view to eventually drafting a *Declaration on the Right to Communicate*. Although not intended for public distribution, this text attracted the attention of ARTICLE 19. The NGO reacted strongly to Hamelink's text stating that it "*seeks to impose a number of wide-ranging and in some cases thoroughly discredited restrictions on freedom of expression*".[14] The polemic between CRIS and ARTICLE 19 began immediately before PrepCom 2 and continued until the end of the second

Preparatory Committee meeting. A session was organized at PrepCom 2 where the two organizations debated their respective visions.[15]

At one point, the difference of opinion around the right to communicate threatened to degenerate into an aggressive political battle. The World Press Freedom Committee, a platform of 44 organizations[16] associated with big media lobby groups, launched an active attack against the CRIS campaign. In their denunciations of the right to communicate, these groups opposed anything that they considered to be a restriction or limitation to the free flow of information, recalling the NWICO debates at UNESCO.

Ideological differences thus created a political situation filled with tensions among civil society members. The large umbrella under which NGOs found themselves at the WSIS included actors associated with the corporate milieu and official institutions as well as far more alternative groups and even some opposed to established power structures and the *status quo*. There was an inherent tension in civil society as debates surrounding the right to communicate are not simply conceptual but are based in a broader political struggle.

Opposition surrounding the right to communicate comes mainly from two directions. Some genuniely oppose the launching of a debate within the United Nations on the definition of a new right to be inserted into the Universal Declaration of Human Rights. This is notably the case of ARTICLE 19:

> ARTICLE 19 endorses, in principle, the idea of an authoritative statement on the right to communicate. However, we are of the view that there is the potential within the framework of existing rights to accommodate the legitimate claims made in the name of the right to communicate. Any elaboration of it must not trench on recognised rights but rather offer an interpretation that expands and strengthens them.[17]

The CRIS position was rather different, and finds its roots in the thinking of the French visionary Jean D'Arcy:

> We wanted to explore in what ways the notion of communication could be taken under the protection of the human rights regime. This exploration is part of a discussion that had begun in 1969 with Jean D'Arcy's famous article on the right to communicate. Jean D'Arcy introduced the right to communicate by saying "the time will come when the Universal Declaration of Human Rights will have to encompass a more extensive right than man's right to information...This is the right of men to communicate". This new approach was motivated by the observation that the provisions in existing human rights law (such as the Universal Declaration of Human Rights or the Covenant on Civil and Political Rights) were inadequate to deal with communication as interactive, two-way traffic and as a process of dialogue.[18]

ARTICLE 19 Law Programme Director Toby Mendel later nuanced the position of his organization:

• Themes Raised by Civil Society •

ARTICLE 19 takes as a starting point the idea that the right to communicate cannot be exercised in a hostile environment and that this implies that States do not impose undue restrictions on content of what may be expressed.[19]

Therefore, according to ARTICLE 19, the right to communicate cannot be applied effectively except in countries that enjoy a relatively high level of democracy; State authoritarianism remains a major obstacle in the realization of such a right.

On the other hand, other actors challenged the very concept of a right to communication insofar as it supposes the establishment of a certain control over the institutions and structures of communication—such as, for example, legislating appropriate thresholds of concentration of media ownership. Some participants in the Media Caucus with close ties to the corporate sector were strongly opposed to any measures that could be perceived as limiting the freedom of media institutions.

The concept of the right to communicate therefore evolved greatly among its partisans in the course of preparations for the first phase of WSIS. In an effort to accommodate those with a valid concern over tampering with the existing international human rights framework, the CRIS campaign modified its approach and moved towards a less formal and more generic term: communication rights.

A right to communicate is now used interchangeably with "communication rights", which is a term less legalistic. The CRIS campaign, for instance, has done that, while moving away from a Right to Communicate that focuses on international law. This is not to deny that international law should make reference to the right to communicate; but rather that it is not a useful or strategic demand at this point and indeed that its pursuit could be counterproductive. The difference might be seen in switching that "everyone should have a Right to Communicate and it should be codified in international Law" to the more colloquial use of rights as in "everyone has a right to communicate and it should therefore be protected and enforced".[20]

Strategic and conceptual reasons explain the semantic transition:

All the trends identified as potentially dangerous to human development can be seen as blocking and limiting people's right to communicate. And it appeals directly to universal rights. Furthermore, the idea that communication, not information or even knowledge, nor simply free speech and freedom of information, should be at the core of reforming media and communication is appealing. The interactive nature of communication, not simply as issuing and receiving information but interacting on matters of substance and thereby setting in motion processes of deepening mutual understanding and of overcoming divisions, is at the heart of media and communication in society.[21]

A conciliation process thus emerged from the conflicts among civil society actors in order to arrive at a shared vision of a human information society where the benefits of open, accessible, fluid and transparent communication would be available to all. From a strict position of opposition, different organizations including CRIS and ARTICLE 19 tried to reconcile their positions in order to move the situation forward. This will for dialogue and to work together was illustrated through the organization of the World Forum on Communication Rights (December 11, 2003 in Geneva),[22] initiated by CRIS and organized jointly with the APC, the Heinrich Böll Foundation, Panos UK, the World Association for Christian Communication, the World Association of Community Radio Broadcasters (AMARC), the People's Communication Charter, the General Intelligence Group, and the WSIS Civil Society Human Rights Caucus.

Intellectual Property Rights, Patents, Trademarks and Public Domain

Civil society at the WSIS became quickly and strongly mobilized around intellectual property rights (IPR) and patent and trademark issues. It developed a discourse focused on the critique of the current IPR regime and the dominant model of development at the WSIS—a market-driven model largely oriented to encouraging private investment. Civil society organizations strongly criticized this approach, assessing that IPR perpetuates inequalities in access, the maintenance of privileges and an inefficient system incapable of redressing the digital divide.

The organizations particularly concerned with the subject organized into a Working Group on Patents, Copyrights and Trademarks. This group rejected the term "intellectual property rights" that "*carries bias and encourages simplistic over generalization*".[23] It preferred the alternative terminology of "intellectual production rights". The criticized term was also frequently replaced by "limited intellectual monopoly", which expresses a negative and limiting connotation. The objective was to challenge the characterization of intellectual property as an intrinsic and absolute right.

Generally speaking, civil society handled the IPR regime using two different approaches. Certain actors lobbied to reform the current system in order to reestablish a balance between private and public interests. Others, such as the Working Group on Patents, Copyrights and Trademarks, focused on a more radical critique of the very concept of IPRs, which are seen as tools for economic domination. In other words, they contested the legitimacy of the in-

• Themes Raised by Civil Society •   87

tellectual protection regime as defined by the World Intellectual Property Organization and the World Trade Organization through TRIPS (Trade-Related Aspects of Intellectual Property Rights). Positions expressed by civil society members at different moments of the preparatory process demonstrate the tension between the two approaches:

The Approach to Reform the IPR Regime	The Approach Contesting the Legitimacy of IPRs
Contribution on Common Vision and Key Principles for the Declaration (PrepCom 2)[24]	**Comments on the Draft non-paper of the President of the WSIS PrepCom on the *Declaration of Principles* (October 30, 2003)[25]**
The global commons, developed as it is by means of public funding and the will of creators, and deriving from our shared physical environment, constitutes a public resource that should not be sold for profits.	'Intellectual property rights' (as distinct from its component of copyright, patents, trademarks etc.) is a relatively recent, industry-driven, concept that attempts to assert that the rights to the use of intellectual products is limited to those granted a temporary monopoly by the state. It suggests others have no rights. In fact, this is precisely the opposite of what is intended with these concepts. The right that all people can use intellectual products is enshrined in the idea of the Public Domain, a legally ancient one and an integral part of all Treaties etc. There are exceptions made to this right, however, the goal of which is to ensure that (while maximum access is maintained for all) mechanisms are also in place to ensure that overall social creativity is also optimised. These exceptions grant a monopoly of use for a period, as a means by which creative effort can be rewarded. It therefore makes no sense to talk of a balance between "intellectual property, on the one hand, and its use, and knowledge sharing, on the other".
The concept of fair use should be protected to maximize the potential of creativity in the public sphere. Non-commercial use of digital contents should be regarded as fair use and thus protected. Authors should be enabled to donate their intellectual contents to the public domain without technological or financial obstacles.	
Global intellectual rights regimes should be reviewed to restore the balance between common interest of sharing knowledge and culture on the one hand and ensure continuing expansion of creation on the other. They should also protect the access to past knowledge, in any new format and media, as part of the global heritage of humanity.	
	The existing paragraph confuses "the protection of intellectual property" with the "granting of temporary monopoly right over the use of intellectual products", resulting in the erroneous suggestion that only such temporary 'owners' have any rights at all.

Table 7. *The Intellectual Property Rights Controversies*

The reform approach was the one gradually adopted by civil society. Although very critical of the current intellectual property rights regime, civil society did not directly contest its foundation but proposed rather profound and significant changes to its nature and scope. Taken from point 2.3.3. of the

Civil Society Declaration, *The Public Domain of Global Knowledge*, the following extract describes the consensus view of civil society.

> Limited intellectual monopolies, also known as intellectual property rights, are granted only for the benefit of society, most notably to encourage creativity and innovation. The benchmark against which they must be reviewed and adjusted regularly is how well they fulfil this purpose. Today, the vast majority of humankind has no access to the public domain of global knowledge, a situation that is contributing to the growth of inequality and exploitation of the poorest peoples and communities. Yet instead of extending and strengthening the global domain, recent developments are restricting information more and more to private hands. Patents are being extended to software (and even to ideas), with the consequent effect of limiting innovation and reinforcing monopolies. Drugs that could save millions of lives are denied to disease sufferers because pharmaceutical companies that hold the patents resist making them available to those countries that cannot pay high prices. Copyright periods have been extended again and again, making them practically indefinite and defeating their original purpose.

Civil society also actively promoted the public domain at the WSIS, fighting against the pro-market position publicized by the other protagonists, and proposing an economic and social development based on socialization and collective solidarity. A fundamental part of civil society's demands was the spread of public knowledge on and off-line, notably through the development of connected libraries, community and university networks.

> Knowledge is the heritage of all humanity. It is an unlimited resource that grows and is enriched as it is shared. Extending and protecting the information in the public domain (global information commons) is a major way of bridging the digital and information divide within and between countries and ensuring conditions for intellectual creativity, technological innovation and participation in the information society.
>
> The personal and public domain knowledge shall be shared between people.
>
> In a democratic society, information and communications are the foundation for transparency, debate and decision-making and for informed choice of an active citizenry.
>
> (...)
>
> Research and academic freedom are keystones of the information society. Academic and public research results should be as far as possible included in the public domain. The public domain plays a crucial role in the creation, evaluation and dissemination of knowledge.[26]

The opening of markets and policies favoring investments would therefore be of little relevance to resolving the problems the Summit addressed, according to this view.

Internet Governance

Launched during PrepCom 2 in February 2003, the Civil Society ICT Global Governance Caucus (previously called the Internet Global Governance Caucus) began its work in April of the same year. The many debates on what positions to take on the subject became particularly lively in the summer of 2003 while the Intersessional Meeting was underway and while civil society was gathering caucus inputs to develop its priorities. The debates quickly centered on the following question: should the WSIS be used as a framework for establishing a constructive critique of the Internet Corporation for Assigned Names and Numbers (ICANN), an organization controlled by the private sector under the patronage of the United States government, and to call on institutional changes within it? Opinion was immediately divided into two concurrent approaches.

The ICANN is a likely target for criticism; it is a private organization registered in California accountable to the US Department of Commerce. Until 2001, it was run by a committee of 19 members representing groups of developers and distributors and by nine directors representing users. A Government Advisory Committee was also open to governments although its recommendations were not binding. According to Wolfgang Kleinwächter, the ICANN has represented "*a kind of Internet government*" for many years by governing the allocation of IP addresses and managing domain names.

Many analysts criticize the US dominance of the organization. Throughout the years, ICANN became very centered on the United States, and the nine directors representing the users did not have the capacity to counterbalance the power of the business sector. Following September 11, 2001, reforms took place that reduced the participation of users and reinforced the participation of governments.

ICANN is said to embody the digital divide and injustice in world communication networks by its geographic location and member composition. On this basis, certain governments seized the WSIS as an opportunity to question the legitimacy of ICANN, criticizing the US government for using obvious privileges in the management of the corporation. Civil society joined the movement of questioning ICANN's legitimacy and started an internal consultation process on what position to adopt on the subject.

The public critique of ICANN was not a given in the prevailing context of the Summit preparatory process. ICANN and the ITU have been fighting a political battle for many years to assume the responsibility for Internet governance. Civil society had to take into consideration that its criticisms of ICANN might risk serving the ITU, an organization closed to the participation of civil society and with a neo-liberal agenda (opening of markets, prevalence of the private sector, deregulation). Civil society equally dreaded the strengthening of government control over Internet management and therefore certain participants maintained a very cautious position on the subject.

The documents produced during this period illustrate the development of discussions within civil society. The following paragraph on the reevaluation of mechanisms for the allocation of Internet names and numbers was present in the *Civil Society Priorities Document* at the Intersessional Meeting in July 2003 but it disappeared in the subsequent version:

> To these ends, the current management of Internet names and numbers and other related mechanisms should be re-examined with the full participation of all stakeholders in light of serving public interests and compatibility with human rights standards.[27]

The Civil Society Declaration emphasized global ICT and communication governance in point 2.4.7. in which it espoused an approach favoring organizational changes:

> Procedurally, decision-making processes must be based on such values as inclusive participation, transparency, and democratic accountability. In particular, institutional reforms are needed to facilitate the full and effective participation of marginalized stakeholders like developing and transitional countries, global civil society organisations, small and medium-sized enterprises, and individual users.
>
> (...)
>
> In light of the relevant controversies in the WSIS process, special attention must be given to improving the global coordination of the Internet's underlying resources. It must be remembered that the Internet is not a singular communications "platform" akin to a public telephone network; it is instead a highly distributed set of protocols, processes, and voluntarily self-associating networks. Accordingly, the Internet cannot be governed effectively by any one organisation or set of interests. An exclusionary intergovernmental model would be especially ill suited to its unique characteristics; only a truly open, multistakeholder, and flexible approach can ensure the Internet's continued growth and transition into a multilingual medium. In parallel, when the conditions for system stability and sound management can be guaranteed, authority over inherently global resources like the root servers should be transferred to a global, multistakeholder entity.
>
> (...)
>
> As a viable first step in this direction, we recommend the establishment of an independent and truly multistakeholder observatory committee to: (1) map and track the most pressing current developments in ICT global governance decision-making; (2) assess and solicit stakeholder input on the conformity of such decision-making with the stated objectives of the WSIS agenda; and (3) report to all stakeholders in the WSIS

•Themes Raised by Civil Society• 91

process on a periodic basis until 2005, at which time a decision could be made on whether to continue or terminate the activity.

However, the governments were not able to reach a consensus on Internet governance. The WSIS entrusted Kofi Annan to form a working group in order to lay the groundwork for the second phase of the Summit in Tunis in 2005.[28] The Working Group on Internet Governance (WGIG) would be one of the key focal points of activity during the second phase of the WSIS.

Gender Issues

Characterized by a high degree of formal and informal involvement, organizations concerned with gender issues were particularly active during the preparatory phase leading up to Geneva. Two civil society groups became the reference points on gender issues: the WSIS Gender Caucus and the NGO Gender Strategies Working Group.

Although gender issues were considered priorities for civil society, certain activists believed that not enough attention was paid to these issues:

> Within civil society as represented in caucuses and families working towards WSIS, the gender approach has found some allies outside the NGO GSWG (Gender Strategies Working Group), yet it has by no means been supported throughout civil society. Within governmental delegations, the Canadians have most consistently championed this approach. In sum, it appears that the majority of representatives—be they from governments, civil society, or business—favor a minimum of references to targeted interventions in limited contexts on behalf of girls and women over gender mainstreaming.[29]

Gender organizations at the WSIS developed a "gender mainstreaming approach", a strategy geared to fight against gender inequalities on particular subjects and issues. It was a question of developing a gender approach that could be applied to all the subject areas discussed.

The Gender Caucus formulated the following demands at the WSIS:

Preparatory processes

1. include gender perspectives in every facet of the Summit—from policy and planning to action, monitoring and evaluation, and also include targets for the participation of women;
2. ensure active participation of gender equality advocates in the preparatory process of the WSIS and the Summit itself to ensure that global ICT policy integrates gender quality goals;
3. include women as leaders and decision-makers in all planning processes for the Summit;

4. facilitate and encourage participation of women as members of national delegations and representatives of civil society and business by setting targets for delegations to include at least 30 percent women including gender and ICT experts;
5. implement an information dissemination campaign that includes a wide range of media such as radio, drama and print, and in a variety of languages, on ICT as a tool for the empowerment of women;
6. commission a study on the interaction between gender equality and ICT, which should include development of a gender equality and ICT baseline, indicators, conceptual tools and case studies on the impact of ICT on achieving gender equality, to inform the dialogue of the Summit.

WSIS Declaration

Include principles in the WSIS Declaration to establish that:

- Information and communication technologies have an important role to play in promoting human development, eradicating poverty and promoting gender equality;

- The benefits of information and communication technologies should be available to all, as recognized in the UN Millennium Declaration and should be promoted through recognition of a Universal Right to Communicate within the international human rights framework. The women's human rights community can serve as a key partner in the development and protection of these rights.

Plan of Action

Develop specific criteria in the WSIS *Plan of Action* to include, for example:

- Programmes to reform decision-making processes in the telecommunications and ICT sector to ensure good governance, greater accountability to all stakeholders and to improve the participation and representation of women and gender equality advocates in all levels of policy making including participation across generations. These programmes should also develop monitoring mechanisms at all levels in the telecommunications and ICT sector to assess the extent of women's greater access and control over resources necessary for their empowerment and to deliver support for capacity building and training that facilitates wide participation by women and gender ICT specialists in policy and decision making;

- Programmes that facilitate women's active participation in the telecommunications and ICT sector through implementing projects that encourage and support entrepreneurship and women's employment, including women's access to international markets;

- Programmes that specifically involve maximizing ICT contribution to the goals of peace, equality and development by developing and encouraging innovative ICT applications that aim to reduce poverty, eliminate HIV/AIDS, promote conflict resolution and peace building, support women's reproductive and productive roles, facilitate education and literacy and reducing violence against women.[30]

The Civil Society Declaration reiterated the concerns expressed by the gender groups and synthesized them in point 2.13, *Gender Justice*.

The Media

Three major media information distribution models may exist and co-exist in democratic systems: private/entrepreneurial, public, and alternative/community.[31] Although the liberalization of State media monopolies had positive impacts on the plurality of information sources available to the public in Western States, it also contributed to reinforcing the development of new private monopolies. With an orientation towards education, citizens and cultures, the public broadcasting model counterbalances the private model that defines the audience primarily as consumers. The idea is to offer quality and accessible information to citizens as part of a public space. Nevertheless, both models are built on hierarchical communication structures; information is produced by a professional elite selected according to criteria of competence and broadcast to citizens who are passive receivers.

The alternative/community model proposes a different approach, one where citizens themselves produce and distribute—in their own tone and manner—information on specific issues of concern to their communities. It therefore becomes possible to address a series of issues ignored by big media in an original way. Many actors believe that community and alternative media contribute significantly to sustainable development. Yet this model remains chronically under-financed despite its major social and democratic promise. Community media protagonists quickly saw the WSIS as an arena to promote this original model on the basis of social and economic development of disadvantaged communities. A number of key community media actors formed a Community Media Caucus, a very different space for expression than the Media Caucus which represented mainly dominant media.

The Media Caucus maintained a double approach at the WSIS, focused on the reaffirmation of the fundamental role of "traditional" mass media—print, television and radio—and on the need to ensure and reinforce freedom of expression and opinion, freedom of the press, and independent media. The debates within the caucus itself were quite numerous because of its heterogeneous makeup:

> The experience of the 'media caucus', that operated under the civil society umbrella with a motley range of actors, was one of the most complex to manage. In reality it was

more of a 'multi-stakeholder' group than a civil society one, since some of the media organizations present were clearly part of the state, others identified with private sector interests, and there were also representatives from UN organizations, in addition to community media groups, journalists' associations, and organizations that defend freedom of expression.

The only issues on which consensus could be reached were the defense and implementation of freedom of expression, access to information including in the digital environment, and the concern that security issues should not affect freedom of expression. It is thus clear that a much broader civil society platform will be needed to defend media reform issues.[32]

Throughout the preparatory process leading up to Geneva, the Media Caucus had to wage a rear-guard action to have traditional media included in the official texts of the WSIS. In fact, traditional media took a back seat to new technologies in the WSIS debates.

This situation was strongly criticized by the Community Media Caucus, which considered it an important failure of the WSIS:

Steve Buckley, president of AMARC, criticized strongly the Information and Communication Technology (ICT)-centered WSIS approach. He stressed that "the Summit's emphasis on ICTs and 'e-strategies' is mainly adequate for elite economies" taking into account that about a third of the world's population has limited or even no access to electricity. He wondered why the WSIS *Declaration of Principles* intends explicitly to "... promote the development goals of the [UN] Millennium Declaration, namely the eradication of extreme poverty and hunger..." without strengthening the most widespread, accessible and cost-effective means of communication. According to Buckley, radio has been proven to be an affordable, decentralized and simple-to-manage medium especially for the poorest and most marginalized communities. At the WSIS about 50 states were willing to recognize the growing importance of community media by mentioning them explicitly in the official documents as tools for poverty reduction and strengthening democratic values. They were encouraged in doing so by the positive experiences of multilateral bodies like the UNDP, UNESCO and the World Bank. Nevertheless, three states—Mexico, El Salvador and China— vetoed any reference to Community Media in the *Declaration of Principles* as well as in the *Plan of Action*.[33]

Throughout the WSIS preparatory process, community media actors deplored the chronic lack of reference to community media in the official texts. There was a rare reference to community media in conjunction with public service media in the *Draft Plan of Action* of August 22, 2003 (WSIS03/PC-3/3-E):

Public service broadcasting and community media have specific and crucial roles to play in ensuring the participation of all in the information society.[34]

This sub-paragraph was subsequently removed from the texts leaving only a modest reference to community media in the official *Plan of Action*:

> Give support to media based in local communities and support projects combining the use of traditional media and new technologies for their role in facilitating the use of local languages, for documenting and preserving local heritage, including landscape and biological diversity, and as a means to reach rural and isolated and nomadic communities.[35]

In the face of the utopianism attached to new information and communication technologies at the WSIS, Steve Buckley recalled that conventional media remain a fundamental strategic choice in a context where, "*according to the United Nations Development Programme (UNDP), two billion people do not have access to electricity and the global population is growing faster than the rate of electrification*".[36]

> For poor people especially those in rural communities, the most widespread and accessible communication technologies remain the traditional media, particularly radio —an oral medium, low cost and receivable by 90 per cent of the world's population. If the WSIS is to contribute effectively to the internationally agreed development goals then it is to the traditional media that it must look first to bridge the communications divide. There is a pressing case to take a fresh look at the traditional media from the perspective of development, eradication of poverty and the rights of poor people.[37]

Positions on community media can be explicitly found in point 2.3.2.2 of the Civil Society Declaration.[38]

At the same time, the Media Caucus advocated for the defense of media freedoms. The positions developed were expressed in the caucus's contribution to the official declaration made at PrepCom 3:[39]

> Freedom of expression, media freedom and editorial independence are central to any conception of an information society.
>
> The guiding overriding principle of WSIS on freedom of expression and media freedom should be Article 19 of the Universal Declaration of Human Rights:
>
> "Everyone has the right to freedom of thought; this right includes freedom to hold opinions without interference and to seek, receive and impart information and ideas through any media and regardless of frontiers."
>
> Article 19 needs to be implemented, for all media regardless of the technologies used.
>
> Security and other considerations should not be allowed to compromise freedom of expression and media freedom.
>
> New information and communication technologies will strengthen the important role of traditional media, such as broadcasting and print press.

Legislation to ensure the participation of all in the information society should:

1. promote and defend the existence and development of free and independent media;
2. encourage pluralism and diversity of media ownership and avoid excessive media concentration;
3. recognize the specific and crucial role of public service broadcasting and community media;
4. transform state-controlled media into editorially independent organizations.

International standards of labor rights and social protections must apply to all media workers.

Formulation of professional and ethical standards in journalism are the responsibility of media professionals themselves.

These elements were reformulated in the Civil Society Declaration in point 2.3.2.1, *The Role of the Media*.

• CHAPTER EIGHT •

Advancing Through the WSIS Preparatory Framework

The different stages that marked the WSIS preparatory process dealt with specific issues related to both the agenda and participation. Certain meetings were more open than others, in regards to both the formal inclusion of non-State actors and the consideration that was given to them. The present section provides a general assessment of the Summit preparation process highlighting some of the most important moments.

The Regional Conferences

The five regional conferences were occasions to grasp different issues and priorities for each region (Africa, Asia-Pacific, Europe and North America, Latin America and the Caribbean, West Asia and the Middle East) and to determine the orientation that different regional delegations wanted to give to the WSIS. Civil society was very present at these meetings with the objective of raising its own issues to different regions.

Each regional conference gave rise to a declaration divided into a statement of principles and priority areas for action that would serve as official inputs to the WSIS. To varying extents and based on specific perspectives, the conferences integrated elements similar to the principles raised by civil society. A survey of the different regional declarations of principles enables us to briefly assess their openness to civil society's principles. The principles relate first of all to what are considered fundamental elements and then to the nine points on the official WSIS agenda: information and communication infrastructure; access to information and knowledge; the role of States, the private sector and civil society in the promotion of ICTs for development; capacity building; security and the creation of a favourable environment; the application of ICTs; cultural and linguistic diversity; local content and media development; the ethical dimension of ICTs.

The degree to which themes dear to civil society were considered varied considerably from one regional conference to another. This is explained by the particular socioeconomic, political and cultural preoccupations of different regions. The African regional conference (Bamako) seems to have been the one with positions and themes closest to civil society's in terms of development and human rights. The extreme precariousness of local populations certainly explains this emphasis. Africa is the continent with the most to gain from a human and solidarity-based information society. The Asia-Pacific conference (Tokyo) equally held a position oriented towards social development. The European/North American conference (Bucharest), reflecting the region where the ICT industry is strongest, focused above all on the development of markets although it remained open to the themes proposed by civil society. The Latin American/ Caribbean (Bavaro) and West Asia/Middle East (Beirut) meetings considered social issues very little, concentrating rather on national and international policies to be adopted.

Generally speaking, the regional conferences concentrated primarily on themes linked to infrastructure development and the establishment of a healthy and favourable environment for investment and installation of ICTs. Bamako and Tokyo were distinct from Beirut, Bavaro and Bucharest based on the themes and issues they prioritized. The place given to the market and the establishment of commercial policies were priorities at the last three regional conferences, yet were relegated to the background at Bamako and Tokyo.

The regional conferences seem to have set the tone for the PrepComs. The principles linked to human development, public information, linguistic and cultural diversity, the integration of youth and gender perspectives, and the democratic role of information and communication, although considered, were not prioritized.

A very liberal approach (in the economic sense of the term) was expressed through the different priority areas for action. Securing investments, network development, public-private partnerships and putting in place business policies compatible on regional and international levels remained the priorities of governments. The African regional conference (Bamako) also set itself apart in this aspect as it concentrated on social needs rather than on the development of private infrastructures.

In general, the market was therefore perceived as the principal motor of social development. Rather than imperatives to be achieved, civil society priorities (development of community access and strong public policies, use of ICTs for disadvantaged groups, development of equitable and effective access to encourage social development) were seen as the intended consequences of investment and market liberalization policies. The technocratic discourse of the ITU seems to have resonated among the governmental delegations to the five regional conferences.

Aside from providing official inputs to the WSIS, the regional conferences thus demonstrated the prevailing political tendencies concerning themes, content and participation. With the exception of Bamako, the role attributed to civil society also seems to have been substantially less important in practice than in principle. Civil society was conceded a role of executor: to be present on the ground, notably with the view of applying public policies with respect to access and effective use of ICTs. The place granted to civil society in the governance of the information society was therefore not very far reaching.

In this respect, the regional conferences served as a bellweather to the PrepComs.

The Preparatory Committees

The three Preparatory Committee meetings were an opportunity for civil society to take stock of the events that marked the WSIS process as it unfolded. Since the Summit's general orientations were largely determined by the PrepComs, they were of strategic importance to all actors, and especially for civil society.

PrepCom 1

Debate on civil society's participation at the Summit began at the first meeting of the Preparatory Committee, where it became obvious that there would be less room for and less attention paid to civil society than it had hoped. As we saw previously, civil society's contributions were given little consideration by the government delegations and played a limited role in the official process. The decision-making process also lacked transparency. The visibility of the NGOs was poor within and outside the Summit, due to the inadequate media coverage of the WSIS. The participants ended up asking themselves whether or not there was any point in lending any further credibility to a UN meeting that was practicing *de facto* exclusion of civil society. The Rules of Procedure adopted at PrepCom 1 allowed for no decision-making power on the part of civil society, which they subordinated to the governmental division.

Under the Rules of Procedure, civil society did not have the right to vote or even to take part in negotiations except at the discretion of the member States. The crucial Subcommittees of the PrepCom were free to decide whether or not they would admit members of civil society, and NGOs and private sector entities shared the same status.

There was much disappointment when the Rules of Procedure on participation were adopted.

> In the most optimistic interpretation, this agreement on rules and modalities for participation represents a variation on established practices, but little in the way of positive innovation.[1]

The WSIS's quiet refusal to include NGOs in the deliberative process prior to the Summit was consistent with the UN's culture. Although civil society's participation in international meetings had recently been broadened, NGOs remain institutionally excluded from the mechanisms through which decisions are made by the governmental or UN actors. At PrepCom 1, the World Summit on the Information Society was actually quite conservative as far as civil society's role was concerned, in contrast to all of the official talk of inclusion.

PrepCom 2

Civil society significantly stepped up its level of involvement in the substantial affairs of the Summit as of Prepcom 2. With little time to prepare, civil society's Content and Themes Working Group was able to produce and distribute important documents that showed that it was capable of formulating positions in a coherent and structured manner.

The impact of the establishment of the Civil Society Bureau at PrepCom 2 has already been mentioned. All criticisms notwithstanding, this development significantly enhanced the capacity of civil society to contribute effectively to the Summit. It also helped to clarify the distinction between the private sector and civil society for the purposes of accreditation although the fact that a number of private entities claimed civil society status in the Bureau sowed a certain amount of confusion.

Prepcom 2 also saw the establishment of a fund to facilitate the participation of members of civil society, and of a new commitment on the part of the Executive Secretariat to increase the transparency of the Secretariat's communication procedures.

A clear pattern of participation was now in place. Generally speaking, Subcommittee and plenary meetings were open to attendance by members of civil society, but intervention was strictly limited to occasional pre-arranged verbal statements. More important, the key intergovernmental working groups remained closed. "Observers groups" were set up alongside these working groups to discuss and produce draft documents on the same issues. The "observers" inputs thus mirrored the official proceedings.

Overall, PrepCom 2 was more open and more transparent than PrepCom 1 and civil society made a number of important procedural gains. However, while the participation process was innovative, it was still anchored in the prevailing logic of previous UN Summits.

One of civil society's major apprehensions at the WSIS was thus the danger that civil society demands would be voiced but not heard.

The Intersessional Meeting

Since work was not moving fast enough to meet the final deadline of December 2003, a so-called "intersessional" meeting was officially convened at the Paris headquarters of UNESCO from July 15 to 18, 2003. The meeting was especially frustrating for civil society insofar as the WSIS appeared to retreat in several key areas of content. As Myriam Horngren of the civil society Communication Rights Caucus (and coordinator of the CRIS campaign) put it,

> The WSIS's latest articulation of communication rights is the narrowest ever imaginable, mostly as the Intersessional ended in Paris on Friday. All references to Human rights have been taken out of the Draft Declaration and with it any reference to communication rights or the right to communicate.
>
> Media are also completely absent, as well as traditional forms of communication. Earlier in the week all this was indeed in the draft. Now we all needed a much shortened document, but it seems that all the cutting back has been done at the expense of human rights, communication rights which include the media and the traditional forms of communication amongst other key issues.
>
> The WSIS is now solely focusing on ITs (without the Cs) and on the digital divide as the only "new information society" issue worth its salt, with the usual framework of market liberalization as the only remedy to solving digital divide (as if it was an illness of itself with no relation to human rights, economic and social issues).[2]

This echoed the opinion expressed by Meryem Marzouki on behalf of civil society at the Intersessional Meeting plenary session on July 18, namely that the lack of openness on the part of the government delegations was worrisome:

> We remind you that in 1948, a little more than 50 years ago, a very strong vision of the future was adopted in the Universal Declaration of Human Rights. Today, with the information and communications techniques we have available to us, we have important tools to help us realize that vision. Rather than express our enthusiasm for

your work, we are obliged to express our fear that you are abandoning that vision, replacing it with technical and technocratic considerations. Instead of progressing towards the full realization of already recognized rights, we are warning you of the danger of real regression.[3]

Communication rights were discussed at the working group level, but the notion was coldly received by the participating government delegations, which actually removed the term from the *Declaration of Principles* because they could not agree on a clear definition. The NGO Gender Strategies Working Group also had to fight to prevent gender issues from disappearing from the draft documents.

But this period also saw a crystallization of civil society's internal organization. Thanks to a broad on-line consultation process, a substantial *Civil Society Priorities Document*[4] was produced and presented to the government delegations at the Intersessional Meeting. Civil society's level of organization and ability to produce content thus increased in effectiveness and quality as the preparatory process unfolded.

PrepCom 3

The third meeting of the Preparatory Committee was the last chance for civil society to have an impact on the WSIS's official texts. It was also a decisive moment in civil society's strategic planning. After over a year and a half of exhausting participation in the process, the time had come for civil society to make crucial choices on its final strategies for the first phase of WSIS.

The deadlock in the negotiations on the most controversial sections of the *Declaration of Principles* led the governmental delegates to set up sector working groups at PrepCom 3. They lost no time putting a cap on civil society participation in these working groups. The negotiations were particularly arduous, and the PrepCom had to suspend its deliberations instead of adjourning at the end of the two weeks scheduled for the discussions; in the end, PrepCom 3 had to meet three times (September 15-26, November 10-14 and December 5, 6 and 9, 2003) before reaching a consensus on the texts to be submitted to the Summit. The governmental delegates had to work up until the last minute to prevent the first phase of the WSIS from being a monumental disaster.

Six official working groups were set up on September 17, 2003 at the beginning of PrepCom 3:

- Right to communicate (chaired by Canada)
- Internet security (chaired by Italy for the European Union)
- Internet governance (chaired by Kenya)
- Enabling environment (chaired by Brazil)

- Cultural identity (chaired by India)
- Media and freedom of expression (chaired by Switzerland)

The modalities of participation established for these working groups essentially whittled the role of civil society for all intents and purposes to a bare minimum. A scant few minutes at the beginning of each working group meeting were allotted for the presentation of civil society's positions before its representatives were asked to leave and the main session continued behind closed doors. Civil society representatives were then allowed back at the end of each session to be told how the negotiations were going. This *modus operandi* was adopted unanimously in a governmental plenary session.

Beginning to feel that they had been enlisted simply to legitimate a non-transparent process over which they had little impact and which threatened to produce results they could not support, many civil society actors began questioning their participation in the WSIS. The discontent was such that the President of the Preparatory Committee, Adama Samassékou, felt it necessary to address the Civil Society Plenary in order to defuse the crisis and avoid a breaking off of relations:

> I wanted to say to you that when I read your report on how things have been going and your analysis, today, I can understand and I agree with your frustration. And I am not saying so to be nice. I am saying so because this is what I deeply feel.
>
> (...)
>
> So, I just have an appeal to make to you in that case: don't throw away the benefit of your own battle. You have sought since the beginning up until now to make your presence felt with the other partners in a constructive approach. Those partners who doubted have understood not only how important you are, but also how productive you can be.
>
> (...)
>
> Dialogue is possible between actors.
>
> Obviously, in any community, there are people who have a propensity for dramatizing, who aren't satisfied if there aren't great difficulties. That's what makes them tick. Unfortunately for these people, I am not one of them. And I hope that there are many of us who will build something great and lasting without any expression of violence. Without accusations of scandal, without attempts to blackmail.
>
> I want this to be clearly understood. This is an appeal that I am making that this matter not overshadow what we have done thus far.[5]

Further behind-the-scenes efforts by the President of the Preparatory Committee managed to avert a total civil society withdrawal but one negative analysis of the process followed another. Civil society integration into the WSIS, once seen as the sign of a new openness, began to be described as a strategic ploy on the part of the Summit's organizers. According to Arne Hintz:

> During the third preparatory conference PrepCom 3, it became increasingly obvious that the opportunities for civil society to participate in the summit process are by no means the result of a gracious gesture by the WSIS organisers. Rather, letting NGOs participate has served to integrate potentially critical voices. A repetition of scenes of street confrontation, as in Seattle, Genoa, or just recently during the G8 summit in Geneva itself, damaging as they would be to publicity efforts, had to be prevented. Thus the "multi-stakeholder approach" has represented a direct response both to the summit protests of the past years and to the lack of legitimacy of large government summits, which had been highlighted by those protests.
>
> At PrepCom 3, even the most cautious points of criticism by the essentially excluded and thereby frustrated NGOs led to sensitive reactions by the WSIS secretariat, the governments, and PrepCom President Samassékou. Attempts to pacify and accommodate civil society were triggered, particularly, by plans for an alternative Civil Society Declaration as that document would have the potential to destroy the carefully nurtured impression of broad civil society support to the official WSIS Declaration.[6]

The civil society structures now began to discuss the limits of the formal procedures and the need to avoid being exploited by the organizers of the WSIS. In a press release dated September 26, 2003, civil society stated that *"if governments continue to exclude our principles, we will not lend legitimacy to the final official WSIS documents."*[7]

The idea was to prevent the WSIS from being incorrectly perceived as an inclusive and multi-partner process, but also to make sure that civil society's criticisms were heard and that civil society was able to present a vision that was more progressive and all-encompassing than the one put forward by the government delegations. This strategy led civil society to announce its partial withdrawal from the official process and explain its new approach to the WSIS in a document issued on November 14, 2003:

> This is the first time that civil society has participated in such a way in a summit preparation process. We have worked very hard to include issues that some did not expect to be included. We have had some successes, while in a number of areas we were not heard or even listened to.
>
> If the governments want to agree, they can agree in 5 minutes. We now have the feeling that there is no political will to agree on a common vision.
>
> Therefore we will now stop giving input to the intergovernmental documents. Our position is that we do not want to endorse documents that represent the lowest common denominator among governments—if there will be anything like that.

We have produced essential benchmarks—our ethical framework—of which we present the latest version today. The governments risk overlooking these key issues in the hair-splitting and compromise of negotiations if they do not take into account our input more seriously.

The current stalemate deepens our belief in the need for the inclusion of all stakeholders in decision-making processes. Where rulers cannot reach consensus, the voices of civil society, communities and citizens can and should provide guidance.

We don't need governments' permission. We take our own responsibility. Someone has to take the lead, if governments won't do it, civil society will do it.

We have now started to draft our own vision document as the result of a two-year, bottom-up, transparent and inclusive online and offline discussion process among civil society groups from all over the world.

We will present our vision at the summit in Geneva in December 2003. We invite all interested parties, from all sectors of society, to join us in open discussion and debate in a true multi-stakeholder process.[8]

So, at PrepCom 3, civil society stabilized its position. After two years of participating in preparations for the WSIS, it took a lucid look at the influence it had had on the process and what its participation had achieved. From then on, civil society would remain present within the walls of the Summit but would refuse to lend its support and legitimacy to the official documents. The governments would have to take sole responsibility for the *Declaration of Principles* and the *Plan of Action* of which they were the sole authors. Civil society had issued a radical critique of the process as it had unfolded, and promised to keep a close eye on the subsequent intergovernmental deliberations. At the same time, it would seek to use the political space provided by the WSIS to refine and promote its own vision of the information society—a pluralistic and communicational society. This vision was finally enunciated in the declaration unanimously adopted by the Civil Society Plenary in Geneva on December 8, 2003.

The Summit

Civil society's problems did not end with the preparatory process. At the December 2003 Summit itself, certain obstacles to a substantial level of participation plagued non-governmental actors. These difficulties ranged from the rejection of civil society's designated speakers by the Executive Secretariat to the lack of access to necessary logistical resources, such as the provision of meeting spaces and computers.

The designation of speakers at the official plenaries was also a sensitive subject, since ideally they would be an objective reflection of the members and opinions of the forces of civil society. At-times laborious negotiations within the civil society structures took place to agree on who was best suited to speak on their behalf. Members of civil society were understandably all the more frustrated, then, when the Secretariat unilaterally decided not to accept the speakers list they submitted on the eve of the Geneva meeting.

According to Sally Burch, co-chair of the Content and Themes Working Group:

> Although the self-organizing mechanisms of civil society provided a list of speakers that was balanced in terms of questions such as geography, gender, topic and prior involvement, that list was largely ignored by the WSIS secretariat.
>
> So when civil society was informed Dec 1st, 2003 by the secretariat who was to speak in its name during the summit, it had to realize that most of the names on that list were unbeknownst to them and even included one mayor of a city, who was apparently to speak in the name of civil society.[9]

This incident was a major irritant for civil society, which pointedly mentioned it in its evaluation of the Summit:

> We had selected our speakers in a fairly transparent and democratic manner before the summit. Then somebody in the ITU just took the list and arbitrarily picked and dropped people. We neither know who took this decision, nor why. But it denied civil society its right to choose who speaks on its behalf and brings its points across. This was especially clear in the opening ceremony.[10]

The Summit was also marred by attempts to whitewash the Tunisian candidacy as host for Phase II of the WSIS. Tunisian nationals accredited as members of civil society entered the area where discussions on human rights violations in Tunisia were taking place, and thousands of copies of the independent daily newspaper *Terra Viva*, which was highly critical toward Tunisia, disappeared in their hands.

These difficulties were well eclipsed by the successes, however. Alongside the official activities in Geneva, the civil society participants organized a number of important events, including The World Forum on Communication Rights, The Community Media Forum, and well-attended sessions on topics such as Media Liberties in the Information Society. A lot of their attention and interest was also drawn by activities and conferences at the World Electronic Media Forum[11] and the ICT for Development platform.[12]

In a more alternative mode, the event *WSIS? WE SEIZE!* took place outside of the Palexpo complex, clearly separating itself (philosophically and geographically) from the Summit-related activities. The organizers of *WE SEIZE!*

rejected the social, political and economic premises of the discussions and debates at the WSIS and proposed a new look at communication and the very foundations of our societies.

> Over the past twelve months, activists and artists from different backgrounds ranging from noborder to indymedia networks, community media activists to grassroots campaigners, have been examining ways to engage the World Summit on the Information Society. This heterogeneous grouping, whose formation began at the European Social Forum in November 2002, operates under the ad hoc banner of 'Geneva03' and comprises people and initiatives with radically different approaches to the WSIS and the issues it raises. Some, for example, will take part in the official process while others won't; some focus on intellectual property while others focus on the struggle for freedom of movement and freedom of communication. All are united by a common understanding that alternative visions and approaches require a strong presence in Geneva during the WSIS.
>
> (...)
>
> It is not our interest to create an 'information-society' which is compatible with the current global system of capitalist society. We want to give answers to the variety of challenging questions raised by the term of 'information-society' that go beyond the horizon of possible answers that could be given by any parts of the official WSIS process.
>
> The groups involved in the preparations of the activities outlined below consider that many of the issues addressed (or failing to be addressed) by the WSIS process are of significant importance to the movement's common social struggles and day-to day activities, though many may not yet be recognized as such.
>
> The first WSIS summit in Geneva presents a timely opportunity to put these issues on the agenda. We have seen how the struggles around access to essential medicines, genetics and free software have transformed patent law from being a shadowy back-road of the law to a matter of public attention. The current litigation and repression strategy conducted by the music industry against p2p users provides a window of opportunity to accomplish something analogous in the field of copyright law.[13]

One of the main activities of *WE SEIZE!*, the *Polimedia Lab*, a space for multimedia creation and innovation, was shut down by the Geneva police. The Lab was later relocated. The Civil Society Plenary condemned the repressive action by the police as an act of censorship in a press release issued on December 12, 2003:

> 12.Dec.03-The Civil Society Plenary, meeting in its final session during the first phase of the UN World Summit on the Information Society (WSIS) taking place from December 10-12 2003 in Geneva, unanimously condemns the undemocratic actions of the Swiss authorities and the Summit organizers in suppressing dissenting and alternative voices.

• Civil Society, Communication and Global Governance •

Over the past three days:

- The Polimedia Lab organized by Geneva '03 Collective, meant to be an open space for participatory communication, was shut down by riot police on Tuesday the 9th of December.
- Printed documents critical of the WSIS and of the media and IT corporate monopolies were confiscated and prevented from being circulated inside the Palexpo, the official venue of the WSIS on 10th December 2003.
- A peaceful demonstration of 50 local and international people at the Gare Cornavin, Geneva, on 12 December 2003, protesting the WSIS and the corporate control of information and supporting community media, was surrounded by about 40 civil police and several vans filled with riot police, and prevented from continuing. Demonstrators were detained, searched, identified and those refusing to be identified were taken to the police station.

These events continue the pattern of political repression that has been a constant feature of public life in Geneva since the G-8 Meeting in early 2003.

We strongly condemn these violations of the right to assemble and freedom of expression that have cast a shadow of hypocrisy over the Summit. [14]

As we thus see, civil society maintained a high level of presence and a structured and coherent involvement both inside and outside the Summit. But what did it all mean?

Integration of Civil Society in a UN Meeting?

The very participation of the NGOs, restricted though it is, raises questions. To what extent is their involvement solicited merely to give the Summit legitimacy? Without the NGOs the emptiness of the windy sermonising might be all the more apparent. On the other hand, the absence of any real decision making intent at WSIS means that there is scarcely a process to 'launder'.

—Alan Toner (2003)[15]

A number of things suggest that there was a real, albeit weak, desire on the part of the organizers of the Summit to include the actors grouped together under the banner of civil society. Although they varied in nature, the difficulties encountered had a great deal to do with the institutional culture of the United Nations. The big UN forums are intergovernmental institutions (and not multi-partner platforms) in which governments remain the repositories of political power. The place civil society is allowed to occupy has always been very limited. The WSIS represented a first hesitant step toward the integration of civil society into the UN process.

The framing of the WSIS as a tripartite Summit was in itself a problem, the heart of which lay in the competing conceptions of multi-partism at international meetings. In the eyes of the member States, the non-State actors are already integrated by virtue of the fact that the consultation has been broadened to them. Civil society, on the other hand, feels alienated by being confined to a role of "consultant" that ties the hands of NGOs and results in the splintering and exploitation of their messages by the governments.

By its very nature, the private sector tends not to seek decision-making powers at such meetings, although it is active in lobbying efforts to increase its influence. Nevertheless, the power of corporations has clearly increased over the last few decades. The end of the polarization between East and West on the international stage and profound changes in the global political economy have redefined the relationships between the private sector and governments. The private sector is now an esteemed "partner" in international affairs.

So, there is no pre-existing agreement between the various parties on what a multi-stakeholder meeting organized by the United Nations should look like. The lack of transparency and the democratic deficit seen at the WSIS were largely determined by the UN's culture regarding organizations that are not extensions of governments.

The structural problems that affected the WSIS were also responsible for the difficulties encountered by civil society. The underfunding of the Summit directly hindered the participation of NGOs (and was especially hard felt by organizations from the South) which struggled to get financial assistance to participate in the event. In fact, those who are generally excluded from the information society were glaringly absent from the Summit itself. The WSIS thus turned out to be a reflection of the same social and participatory divides it criticized. As it was incapable of offering the necessary resources for acceptable civil society participation, the WSIS marginalized those with most at stake in the official agenda: organizations from the Global South, emerging countries, and participants from States in economic and political transition. The weak financing of the WSIS directly impacted the geographic, demographic and ethnic balance among WSIS civil society representatives. This imbalance was present as of the NGO preparatory consultations organized by UNESCO in February 2002. According to one observer:

> It is hardly surprising to find that most of the NGOs attending either have offices in UNESCO itself, or else are located in adjoining European countries. The result is an extreme white Western (and middle-aged—or even older—male) bias in representation. There is also a limitation posed by the kinds of NGO with which UNESCO has 'formal relationships'. These tend to be 'professional', as opposed to 'civil rights' or 'activist' in orientation.[16]

In this context, the good will of the WSIS Executive Secretariat could only have a limited impact in the face of States that were little inclined, if at all, to allow any expansion of the official role of entities which were not extensions of themselves. The Secretariat was more administrative than policy-oriented as a structure, its primary responsibility being to ensure that work proceeded as planned by enforcing the will expressed by the governmental delegates, particularly as regards the participation of non-State actors. Although its actions themselves were limited, the Secretariat was generally seen as acting in good faith. The belated creation of a Civil Society Facilitation Fund expressed this willingness (albeit limited) to include civil society in the official process. Civil society considered the President of the Preparatory Committee, Adama Samassékou, to be an ally as he actively supported the creation of the Civil Society Bureau and was generally open to the proposals and needs expressed by civil society. Samassékou personally attended many formal and informal meetings with civil society members. The President of the Preparatory Committee quickly stepped in as moderator between civil society and governmental delegations seeking to diffuse crises by stressing the need for dialogue and patience. Civil society representatives also had informal meetings with the ITU's Secretary General, Yoshio Utsumi, on many occasions.

But the Civil Society Division did not have the resources it needed to carry out its mandate. The Executive Secretariat had to deal with civil society demanding a more significant place at the WSIS on the one hand and institutions that were disinclined to contribute to the achievement of the political objectives of the non-State actors on the other. The result was a climate of mistrust between the organizers of the WSIS and civil society, raising questions as to how real the interest in inclusion of non-State actors actually was.

The experience of the WSIS showed that governments are still not open in a comprehensively meaningful way to a more complete integration of civil society actors on the international stage. Negotiation is an activity that rests on relative positions of strength and governments are not very excited about the idea of giving any ground to organizations that are highly critical of them. Confining civil society to an observer role unquestionably provides governments with a dual benefit by legitimizing an event in which civil society is participating and protecting their prerogative to make decisions. Seen this way, it is true that civil society helped make the WSIS look better than it may have deserved—which is precisely the role that the civil society organizations refused to play in the end.

PART THREE
Outcomes of the First Phase of the WSIS

The end of the first phase of the World Summit on the Information Society on December 12, 2003 was the culmination of a long process that started five years earlier at the ITU Plenipotentiary Conference. Tremendous efforts had been invested in the undertaking. The Summit's stakeholders—civil society, the private sector, UN agencies and governments of UN member States—all struggled to affect the outcomes on the basis of their respective visions and interests. Any assessment of the WSIS will therefore necessarily be complex and depend to a large extent on the analyst's perspective.

Without attempting to be exhaustive, this section will present a three-part assessment of the WSIS based on an analysis of the Summit's results. First, we will focus on the official outcomes: the completion of the negotiations, the themes selected for consideration in Geneva and those put off until the second phase in Tunis. We will then present a critical analysis of civil society's participation in the process, the official results it obtained and its independent achievements. The third part of our assessment will be a more general discussion of the Summit, conceptualized as an arena in which different models of international governance and communication are being confronted.

PART THREE

Outcomes of the First Phase of the WSIS

• CHAPTER NINE •

The Official Outcomes of the WSIS

> We are firmly convinced that we are collectively entering a new era of enormous potential, that of the information society and expanded human communication. In this emerging society, information and knowledge can be produced, exchanged, shared and communicated through all the networks of the world. All individuals can soon, if we take the necessary actions, together build a new information society based on shared knowledge and founded on global solidarity and a better mutual understanding between peoples and nations. We trust that these measures will open the way to the future development of a true knowledge society.
> —WSIS Declaration of Principles (2003)[1]

With over 11,000 officially registered participants, the first phase of the WSIS fit well into the mould of UN global gatherings. It was also a costly event at over 11.8 million Swiss francs, of which governments contributed roughly four million[2] (USD $3.4 million). The Palexpo centre in Geneva was occupied by preparations for weeks before the Summit began.

The WSIS was however the first global Summit devoted to communication governance and policy issues. The dawning of a digital "revolution" in the words of Summit organizers,[3] with its attendant social, political, economic and cultural impacts, was considered important enough to persuade the International Telecommunication Union to take on the task of organizing such a large-scale event.

Yet the WSIS failed to attract a great deal of attention in the mainstream media, which apparently felt that it lacked the thematic appeal of some of the UN's earlier global meetings (Earth Summit, Rio de Janeiro, 1992; Human Rights Summit, Vienna, 1993; World Conference on Women, Beijing, 1995; World Summit on Sustainable Development, Johannesburg, 2002). It was particularly difficult to get the WSIS's themes and issues across in clear terms to the media and the public. The organizers had to work tirelessly to present the WSIS in a less technology-centred light and to position it in a context of social and human development. The correlations between poverty, social and economic exclusion, human rights and information and communication technologies were not adequately and clearly established. Ultimately, WSIS went relatively unnoticed by the peoples of the world, unable to find its place on the list of concerns in world public opinion.

As far as public opinion was concerned, the WSIS was in many respects a "silent Summit", unadorned by the pomp and circumstance that usually surrounds major international events. It has even been suggested that part of the reason for this was that civil society was largely on the inside rather than outside in the streets.

Nor did the WSIS register very high on the political scale. It was attended by less than fifty Heads of State, mostly from Africa and the Arab world, and it seems that high-ranking Western politicians had more important things to do.

The six months leading up to the first phase of the WSIS were an ordeal for the diplomats sent to negotiate at the Preparatory Committee meetings. The talks were so difficult right up to the eve of the Summit that some particularly controversial items were put off to the second phase in Tunis in 2005. In fact, after the third meeting of the Preparatory Committee, *La Tribune de Genève* gave the Summit little chance of success:

> Is there anyone who is still prepared to bet on the success of World Summit on the Information Society (WSIS) that is supposed to take place in December? At the end of the "last" two weeks of negotiations before the main conference itself, the only thing that key people in charge could report on yesterday were the disagreements.[4]

Le Temps came out with a similar analysis less than a month before the Summit was held:

> Over the last two and a half years, these fine ambitions have run up against a series of obstacles: an international agenda dominated by security concerns, trade block tensions between the North and certain powers in the South, and finally, the reticence of a number of countries that balk at any encroachment on their prerogatives. The latter happen to be the very countries that have the most to fear from the new communication technologies.
>
> Behind the current disagreements (over financing, human rights and Internet governance), governments are actually raising the question of their mutual solidarity, of a minimum level of common human standards and, finally, of their sovereignty, in other words, of their legitimacy.[5]

More specifically, four points remained especially problematic at the end of the negotiations: human rights and the freedom of expression, intellectual property rights, financing of the information society, and Internet governance.

A number of governments found themselves repeatedly at the centre of controversies. Those hostile to the participation of civil society like Pakistan,

Iran, Russia and China stuck to positions which others found hard to accept. China in particular objected to any inclusion of human rights standards and reference to media. With the support of a number of governments, most notably Pakistan and Russia, China rejected "the free circulation of information" in favour of regulation subject to "national legislation", thereby refusing to subordinate its action to international standards. The groups advocating freedom of expression were particularly critical of this stance that unequivocally grants States the prerogative to define the conditions of pluralism and independence in the media.[6] Russia maintained that securing information was a military responsibility until PrepCom 3 resumed, at which time it accepted the less aggressive expression "information security" proposed by the United States.

Human rights issues remained problematic until the final hours leading up to the Summit. After having been put on hold, and almost disappearing during part of the preparatory process, references to human rights were finally included in the official documents. It was only after a long and laborious lobbying effort that the human rights organizations were able to heave a sigh of relief, with the reaffirmation and consolidation of Article 19 of the Universal Declaration of Human Rights, which states that

> Everyone has the right to freedom of opinion and expression; this right includes freedom to hold opinions without interference and to seek, receive and impart information and ideas through any media and regardless of frontiers.

The existing intellectual property rights system was the subject of a major debate between certain developing countries and the developed North. Brazil disagreed with the Northern countries on the wording of provisions to protect intellectual property rights. There was also a long and drawn-out argument over the wording of paragraphs on free, open source and proprietary software.

Senegalese President Abdoulaye Wade put forward an initiative on financing for the information society that failed to win unanimous approval from the representatives of the States. While widely supported by the developing countries and civil society, President Wade's proposal to create a "painless" Digital Solidarity Fund, based on voluntary contributions and a kind of "tax" on ICT products, was not to the liking of the Northern countries. The United States, the European Union, Canada and Japan spoke out against the idea, preferring the existing financing mechanisms. The result was a watered-down compromise: States interested in participating would be able to do so on a voluntary basis. Still, President Wade did not leave empty-handed: he won the support of the cities (Lyon and Geneva contributed 600,000 euros to launch the Fund) and of many non-governmental organizations. Further discussion of financing was put off until Tunis.

Consequently, there was no clear mechanism for the funding of the resolutions adopted in the official *Declaration of Principles* and the *Plan of Action*. This was clearly a major failure of the Summit and a flagrant sign of the political unwillingness of the Heads of State to adopt the measures needed to translate the principles adopted at the WSIS into reality. Given that the President of the Preparatory Committees Adama Samassékou had already said that he agreed that *"the funding of concrete actions will be the first criterion for the success of the Summit"*,[7] this set-back was all the more revealing of what one can really expect of efforts to carry out the mandates adopted by the WSIS.

The present Internet governance system was also challenged by some developing countries, which have questioned the *status quo* that gives the US control. This thorny subject was also put off until Tunis, giving the WSIS organizers more time to bring the sides closer together. The Summit appealed as follows to Kofi Annan:

> We ask the Secretary-General of the United Nations to set up a working group on Internet governance, in an open and inclusive process that ensures a mechanism for the full and active participation of governments, the private sector and civil society from both developing and developed countries, involving relevant intergovernmental and international organizations and forums, to investigate and make proposals for action, as appropriate, on the governance of Internet by 2005.[8]

Considering that the two most fundamental issues—Internet governance and the financing of the information society—were thus put off until Tunis, the first phase of WSIS clearly accomplished little in the line of concrete proposals. But while the States themselves must have had mixed feelings about the outcome of the Geneva phase of the Summit, they still saw the WSIS as having made it possible to work out a common language with which the new information society could be approached and some common definitions of the underlying issues.

The *Declaration of Principles* defines the following "key principles" of *"an information society for all"*:

> We are resolute in our quest to ensure that everyone can benefit from the opportunities that ICTs can offer. We agree that to meet these challenges, all stakeholders should work together to: improve access to information and communication infrastructure and technologies as well as to information and knowledge; build capacity; increase confidence and security in the use of ICTs; create an enabling environment at all levels; develop and widen ICT applications; foster and respect cultural diversity; recognize the role of the media; address the ethical dimensions of the Information Society; and encourage international and regional cooperation. We agree that these are the key principles for building an inclusive Information Society.[9]

• The Official Outcomes of the WSIS •					117

Although it is seriously weakened by the absence of concrete funding measures, the *Plan of Action* supports the following actions to concretize the vision outlined in the *Declaration of Principles*:

- to connect villages with ICTs and establish community access points;
- to connect universities, colleges, secondary schools and primary schools with ICTs;
- to connect scientific and research centres with ICTs;
- to connect public libraries, cultural centres, museums, post offices and archives with ICTs;
- to connect health centres and hospitals with ICTs;
- to connect all local and central government departments and establish Websites and email addresses;
- to adapt all primary and secondary school curricula to meet the challenges of the information society, taking into account national circumstances;
- to ensure that all of the world's population have access to television and radio services;
- to encourage the development of content and to put in place technical conditions in order to facilitate the presence and use of all world languages on the Internet;
- to ensure that more than half the world's inhabitants have access to ICTs within their reach.

Beyond this nominally generous set of principles and plan of action, the official documents produced at the WSIS reflect a specific conception of the world in the issues they conceptualize and the procedures for intervention they propose. They form a framework built around a dominant model which legitimizes certain types of action and focuses on certain themes, and which is challenged as biased by many authors.

Marita Moll and Leslie Regan Shade, for example, contributed a very relevant critique of the WSIS in an article entitled *Vision impossible? The World Summit on the Information Society*:[10]

> Given the devolution of power from UN agencies, some have questioned whether any meaningful discussion can be held at WSIS. The Draft Principles and Agenda for Action are extraordinarily ICT focused thus "resuscitating a ruse reminiscent of the heights of the 'dot com' folly: addition of prefix 'e-' to any given area of human activity to cast it as an 'ICT issue' (e-administration, e-learning and so on)". This technocratic discourse, although not unusual given its predominance in other policy discussions on ICTs for development, lends itself to top-down decision making rather than collabora-

tive processes. And, the focus on the digital divide supports industry imperatives that market forces are the only way to provide technological resources. This emphasis obscures important issues related to the social infrastructure, such as increasing educational resources in support of literacy, and even to providing viable physical resources in communities.

Absent from the WSIS discourse are debates about whether or not ICTs are appropriate tools for development, a contemporary debate that has been rehashed with the activities surrounding the DOT Force (Digital Opportunities Task Force), the G8 initiative to "ameliorate the digital divide" in developing countries. As with the DOT Force, official WSIS discourse is relatively uncritical; as in previous debates on strengthening communication systems for developing countries, current discussions are concerned with the "how and when to 'connect' communities in the South instead of with the why, who, under what conditions, and with what implications". And, similar to the DOT Force pronouncements, WSIS reveals its allegiance to the modernization paradigm, wherein technology is equated with development.

AMARC President Steve Buckley concurred with this critique of the technocentric utopia that has been the dominant model at the WSIS:

> It should be obvious to anyone living outside a fictional Internet utopia that the poor people need clean water more than they need fast connectivity even though access to good information can help make water clean.

Others have argued compellingly that giving universal access to the Internet will cost a lot and accomplish little. Bill Gates, speaking in October 2000 at a Seattle conference on the "digital dividend", famously argued that investment in health and literacy is more important for poor people than providing access to PCs and the Internet.

Charles Kenny, an economist with the World Bank, has estimated that the worldwide subsidy needed for everyone living on $1 a day to get one hour of access a week might reach $75 billion—considerably more than the global total aid flows each year.[11]

Any narrowing of the digital divide appears unlikely in the face of estimates that an additional USD $88 billion a year would be needed to achieve the objectives laid out by the Millennium Summit.

So the World Summit on the Information Society did not give itself the means to achieve its ends and failed to distinguish itself from other recent UN global gatherings such as the Millennium Summit in 2000 and the World Summit on Sustainable Development in 2002 in that it too consumed a great deal of time, energy and money just to produce texts that are likely to be acted upon little if at all.

• CHAPTER TEN •

The Achievements of Civil Society

> Our participation in the WSIS process has been intense, in both human and financial terms, and many people of course have been unable to participate, notably from the poorest countries. Despite these constraints, civil society has produced many contributions to this meeting. We have offered diverse and practical recommendations.
>
> —Meryem Marzouki (2003)[1]

As the first UN Summit to which civil society was invited as an official actor, the WSIS was seen by many as a wonderful opportunity to get involved in international governance and to bring about changes in its rules. They soon lost their illusions as it was quickly and clearly demonstrated to the NGOs and other members of civil society that their seats at the negotiating tables were not guaranteed.

The first meeting of the Preparatory Committee had the effect of a cold shower on civil society, but that did not prevent it from organizing itself into coherent and functional structures that greatly increased its credibility in the eyes of the other actors. The concrete advances made as of PrepCom 2 were mainly of two orders. Firstly, the creation of a Civil Society Bureau was a major political victory. For the first time in the history of the United Nations, infrastructures were put into place to involve civil society, institutionalizing its participation in the Summit and creating a precedent for meetings to come. Secondly, but no less importantly, civil society succeeded in maintaining a high level of cohesion throughout the preparatory process and was able to pool its members' resources to produce strong, high quality consensus-based documents. Cooperation among the actors, which culminated in the production, adoption and circulation of the declaration *Shaping Information Societies for Human Needs*, was the key to civil society's success at the Summit.

As we have seen, civil society actors had to fight a constant uphill battle to maintain a minimally acceptable level of participation in the Summit's official proceedings. The Preparatory Committee meetings were closed or open depending on the sole discretion of the government delegations, and civil society's many contributions had little real impact on their negotiations. An analysis done in September 2003 by a group of civil society volunteers showed that, as of that time, 60% of civil society proposals had been completely rejected, 15% were "*more or less taken into account*" and 25% appeared in some form in the most recent version of the documents.[2]

As the Summit approached, Bruce Girard and Sean Ó Siochru of the CRIS campaign offered a mixed assessment of the results of civil society's participation. The promises of inclusion, proclaimed loudly and clearly by the event's organizers, had for the most part not been kept.

> So while promises of a new type of summit were undoubtedly sincerely meant, the reality falls far short of these. Most of the hopes expressed at the Paris meetings were unfilled.
>
> - There is little or nothing in the way of new modalities for participation for civil society, and indeed existing modalities have not been optimised. The future holds, at best, further uncertainty.
> - No protocol has ever been issued outlining and confirming the transparency of the entire WSIS process, from Bureau to accreditation procedures.
> - Civil society has been offered no representation on the Bureau, though the creation of a Civil Society Bureau might yet facilitate more meaningful interaction.
> - There is no concerted process to stimulate civil society interaction and participation in the Summit, beyond the minimum implemented by the poorly funded CSD; ideas such as civil society "animators" have not been realised for want of resources.
> - Funding for civil society has also remained sporadic and arbitrary, and a dedicated fund has yet to be established, though this might improve as the Summit approaches.[3]

Civil society attempted to influence the outcome of negotiations through official and institutionalized action as well as informal lobbying. More formally, civil society intervened through its officially distributed declarations, its involvement in the round tables and short declarations in the plenary sessions or Subcommittee meetings. However, civil society got its best results from its more informal lobbying of the inter-governmental working groups. These groups, which were set up at the Intersessional Meeting in Paris (July 2003), were able to reconcile some countries with differing views and worked to draft consensus positions for presentation to the governmental plenary. Despite the fact that civil society was formally excluded from the deliberations of the working groups, several actors were able to influence the negotiations by providing the delegates with arguments based on their respective areas of expertise. All things considered, these efforts produced some interesting results.

To be effective, according to Sean Ó Siochru, civil society had to develop both its internal interactions and external networking activities.[4] Internal interactions, needed to develop positions and arrive at a consensus, were accompanied by the building of networks with WSIS government delegations considered to be "friendly". This is how civil society was able to influence the official negotiations in the targeted areas.

Civil Society's Assessment of the WSIS Process

As committed as they were to the participatory process, the civil society actors remained highly critical of the content officially produced by the State representatives throughout the preparatory process. Ultimately, civil society responded to the lack of consideration of the inputs of its members, and the development of positions contrary to those it was advocating, by withdrawing its support for the official texts and embarking on a process to draft an alternative to the intergovernmental *Declaration of Principles*.

Sally Burch, co-chair of the Civil Society Content and Themes Working Group, summed up her take on the first phase of the WSIS in an article in the journal *Media Development*:[5]

> Most CSOs (civil society organizations) concur, nonetheless, that overall the official Declaration and *Plan of Action* express tepid commitments and show feeble political will of governments to address the fundamental issues. In particular, there was no decision concerning funding for telecommunications development in developing countries, and no agreement on broadening participation in Internet governance mechanisms, both of which have been remitted to task forces and postponed to the Tunis phase of the Summit.
>
> (...)
>
> It took over a year for governments to agree to mention even the Universal Declaration of Human Rights (UDHR) as a basis for the information society. The full quote of Article 19, on freedom of expression, was also hotly debated, and only accepted by some countries when accompanied by a qualifying clause that could open the door to national exceptions.
>
> The reference to 'the right to communicate', included in initial drafts of the Declaration, was subsequently eliminated from the official documents, as there was no consensus on its interpretation. For some, it implies universal access to telecommunications (and as such, interestingly, was supported by both ITU Secretary-General Yoshio Utsumi and by Kofi Annan, UN Secretary-General). For others, such as the CRIS Campaign, it embraces the full range of existing rights associated with communication, but also implies the need to consecrate new rights, that are becoming necessary in the present communications context.

But some actors oppose the term because of its association with the battles around NWICO in UNESCO in the 1980s, and consider the existing framework of Article 19 of the Universal Declaration of Human Rights to be sufficient, though not adequately enforced. It is significant, all the same, that the debate on communication rights has been renewed in the WSIS context and that there is increasing recognition that the existing framework of rights needs to be reinforced and broadened. The World Forum on Communication Rights, organized by CRIS and others during the Summit in December, was one space where this debate has continued.

(...)

Some small advances were achieved by civil society at the WSIS in relation to a number of such issues, although many of them might be more accurately described as 'damage control', that is, avoiding inclusion of the most unacceptable language, which even so could not always be averted.

(...)

In summary, most actors in the WSIS process will be able to find language in the final documents that they can use as support for their agendas, and to leverage support from governments and international institutions. But many other issues are absent or inadequately dealt with and overall there is little coherence. The Civil Society Declaration is a much more coherent document that—while there is room for further development and refinement of the proposals—will be a reference point, not only for the next phase of the WSIS but also for many organizations concerned with these issues in other spheres.

A number of participants and observers have expressed positions that are complementary to Sally Burch's appraisal. Some, like Georg C. F. Greve, President of the Free Software Foundation Europe, saw the official documents as deficient but moving in the right direction:

> One can say that the governmental documents fall short of the essential benchmarks of civil society in all considered aspects. But one can also say that they have in most cases made progress and moved in the right direction.[6]

Others were less charitable. The Choike portal deplored what it saw as a pro-market and technology-centred vision in the official texts, a vision that leaves little room for the egalitarian development of the information society:

> Although part of the content promoted by civil society was included in the official documents—for example, references to the defense of human rights—a detailed analysis of the text reveals a vision of technology promoted by commercial interests, in contrast to the conception held by the majority of civil society actors of technology as a tool for egalitarian development. Powerful pressure groups, such as the corporate media, left their mark on the documents, which locate other more democratizing forms of communication, such as community-based media, on the margins of the information society.[7]

CRIS spokesperson Sean Ó Siochru was also critical of the outcome of the first phase of the WSIS. On many points, he wrote, the Summit has done little more than sanction the *status quo* by refusing to take up fundamental issues and limiting itself to timid and inadequate proposals for concrete action. But beyond any expressions of disappointment over the tangible results one might have expected of the event, civil society situated itself in another paradigm and articulated new concepts on life in society and communication among human beings. The idea is not "simply" to build a more equitable information society, but to work toward the development of a communication society, to re-examine the structures of power and domination as expressed, for example, through the mainstream media, and to develop common and universal knowledge.

> As noted, civil society ultimately decided to withdraw from (though not to oppose) the official WSIS process, in the special PrepCom 3A convened in mid November 2003, less than a month before the Summit itself. The reasons given are instructive, as they illustrate not just a difference regarding how to treat the items included in the WSIS agenda, but a desire for a wider agenda. (...)
>
> These essential benchmarks, delineating the main differences between the civil society and the intergovernmental position, included the important areas of difference within the 'information society' agenda, arguing that in human rights, poverty reduction, sustainable development and social justice should be more securely built in. But they also included all the key 'communications society' issues such as the role of "communications media and information technologies" in promoting diversity of culture and language, and of "editorially independent public service media organizations"; and the need to encourage "pluralism and diversity of media ownership... to avoid excessive media concentration". "Copyrights or patents", it noted, "are granted only for the benefit of society, most notably to encourage creativity and innovation." All of these were hugely watered down in the official WSIS agenda, or off the agenda altogether.[8]

Independently of the Summit's official outcomes, the greatest accomplishments of civil society remain the achievement of a high level of coordination among its constituent entities, the development of networks, expertise and common projects, the exchange of ideas and methods and the articulation of an alternative discourse that benefits from the spotlight of a high-level UN meeting.

It is in this sense that the declaration unanimously adopted at the Civil Society Plenary session on December 8 is more than a political statement of principles. It is also the concretization of a long process that could profoundly change the way non-State actors perceive the range of means of action available to them at such events. This is a success which reinforces civil society in the new roles that have opened up for it due to the significant changes underway in international governance, changes which will inevitably reach far beyond the WSIS.

• CHAPTER ELEVEN •

The WSIS as a Model of Communication Governance

The World Summit on the Information Society can be seen as a space of converging trends that are having a structural impact on both international governance and communication. The dominant international decision-making models and normative systems, sets of procedures and forms of participation are being reformulated. Many observers now view the WSIS as an arena in which embryonic transformations in international governance have manifested themselves, as a "laboratory" where the first faltering steps of a new form of governance are being tested. This thesis is supported by a number of significant developments—the historical precedent for the UN of a multi-partner Executive Secretariat, the extensive integration of civil society and the private sector in the official process, the creation of a Civil Society Bureau, the credibility that the quality of civil society's independent action has won for it. All of these developments are fostering a second level of thinking on the WSIS that shifts its attention away from the substantive issues of the Summit, which we have focused on here, to consider the WSIS itself as the expression of new trends in international governance. The WSIS is thus considered as a place where a new distribution of power is taking place among different actors.

The WSIS was also the meeting place of divergent analytical approaches to international communication, of different paradigms competing over the definition of the information society and its global dimensions. In this sense, the struggles around WSIS have been conceptual and discursive as well as political and strategic. The WSIS reformulated positions already debated in other major international forums; in particular, it rearticulated and updated several themes of the debate on the New World Information and Communication Order that took place a good quarter century earlier.

The difference between the context of the NWICO debate and that of the WSIS is of course that global politics have changed dramatically since the 1970s. But despite that obvious difference, and despite the even more dramatic developments in communication, the issues remain remarkably familiar. But not the players. New global, transnational and internationalist civil society actors have appeared on the scene and are showing that they are capable of acting politically as a responsible, organized and coherent force, respectful of the autonomy and difference of its constituent parts. New communication technologies have been critical to the emergence of that force. They are also strategically vital to its further development. At the WSIS, these technologies were also the object of discussion, negotiation and struggles for power. Therein lies what we are referring to as a convergence of trends that is shaping the emergence of a new paradigm of global governance.

The WSIS and Global Governance

Governance is a polysemic concept defined more or less broadly depending on the use to which it is put. There is no conceptual consensus on the notion of governance—and this book is surely not the place to initiate the much-needed debate on the question. That said, the main international institutions each has its own conception of governance. For example, the World Bank espouses the following traditional definition:

> Governance is the manner in which power is exercised in the management of a country's economic and social resources for development.[1]

The World Bank's definition thus refers to three fundamental aspects of governance: (1) the form of political system; (2) the process whereby authority is exercised in the management of social and economic resources for the development of a country; (3) the ability of government to designate, formulate and implement policies and discharge functions.

The process described by the World Bank is vertical and hierarchical. Governance is largely the exclusive domain of the government apparatus; the definition makes provision neither for interactions nor for ways in which power is exercised nor for interrelations between different structures and levels of application. It also fails to mention the actors involved. In other words, it is not a satisfactory definition because it is incapable of fully expressing the reality to which it applies.

The United Nations Development Programme (UNDP) proposes quite a different definition of governance:

• The WSIS as a Model of Communication Governance • 127

Governance—the exercise of political, economic and administrative authority in the management of a country's affairs at all levels. Governance is a neutral concept comprising the complex mechanisms, processes, relationships and institutions through which citizens and groups articulate their interests, exercise their rights and obligations and mediate their differences.[2]

This much more dynamic definition considers the political, economic and administrative process, takes into account the plurality of actors involved in the governance process and the complexity of the interactive multi-level mechanisms they are engaged in, and considers the relations, processes and institutions involved. The actors are invited to articulate their interests, exercise their rights and obligations and mediate their differences.

The choice of these two competing definitions is not gratuitous; it expresses a shift from a traditional and hierarchical vision of global governance to a process of dynamic integration that is opening up to new actors and taking place at various levels.

Let us look now at some of the authors who see in the World Summit on the Information Society a manifestation of profound changes in the global governance system.

Building on the analyses of Baylis and Smith (1997) and Nye and Donahue (2000), professors Claudia Padovani and Arjuna Tuzzi of the University of Padova state:

We assume that new forms of politics are possibly emerging, with new actors being more and more recognized as legitimate on the global scene; international intergovernmental organizations, private entities and civil society organizations.[3]

Starting from the premise that the WSIS *"is a process that offers insights for a better understanding of governance in the global sphere"*, the authors call for the establishment of *"alternative modalities for the 'authoritative allocation of resources' since the changing nature of politics now requires (and implies) new modes of decision making and political participation"*.

The public space is undergoing profound changes, as the multiplication of international NGOs (Brown et al. 2000) as well as the emergence of so-called "no-global" movements and trans-national "protest politics" demonstrate (Tarro 1999; Della Porta, Kriesi 1998). Such changes are not exclusively but deeply grounded in the adoption and diffusion of new information technologies and in the role played by communication and information sharing in civil society trans-national activities (Keck & Sikking 1999). The public space is becoming truly trans-national and its 'citizens' are now asking for institutional mechanisms and normative guarantees that allow them to participate meaningfully in developing not just norms and visions, but also concrete structures of effective global governance (Nye, Donahue 2000). The more political issues are global in scope and are aggregated at the global level with the contribution of trans-national actors, the more responses must be found and policies elaborated through negotiation mechanisms capable of offering representation and voice to the different interests at stake.[4]

Rosenau (1999) sees the major changes affecting the international system as four-fold:

- the shifting in the location of authority towards supra and extra-national forums;
- the emergence of a trans-national civil society;
- the re-orientation of intellectual, political and economic elites;
- the emergence of a globally oriented epistemic elite.

Padovani and Tuzzi take Rosenau's observations and develop the connection between the WSIS and global governance as follows:

> (1) We can think of the Summit as a 'shift in the location of authority' both for the fact that there is a recognized need to face challenges posed by and to societal transformation at the highest political level and for the fact that a number of supra-national political instances (IGOs, programmes, agents) involved in the attempt to regulate such changes converge in this process; (2) we see the 'emerging transnational civil society' mastering its capacity to become part of a high level political process building on former experiences, not only acting as an observer or submitting contributions, but also influencing in different ways the development of the process and suggesting ways for a better involvement of non-governmental actors; (3) we can see the WSIS as an opportunity for the 'intellectual, political, and economic elites' to debate respective orientations relating to the transformation of information economies and knowledge societies; (4) finally we witness a gathering around such process, precisely of members of that epistemic elite defined by Rosenau as the 'technicians, experts in knowledge', trying to bring their perspectives and contribution of a 'common vision of the information society'.[5]

The inclusion of non-traditional actors at the WSIS, intended and organized by the initiators of the event when the preparatory process began, is an illustration of the changes that are taking place in international politics.

So, for Padovani and Tuzzi, the concept of governance can no longer be applied exclusively to intergovernmental relations; we are witnessing the emergence of a global public space that calls upon new actors and new modes of management as it takes shape.

This analysis is largely shared by us. In an article published shortly after Geneva, Marc Raboy, who was actively involved in the process leading up to the WSIS both as a researcher and as a member of the CRIS campaign, wrote:

> Regardless how one looks at it, the World Summit on the Information Society undeniably opens a new phase in global communication governance and governance generally. (...) The global governance environment in communication (as is much

everything else) is based on the interaction and interdependence of a wide array of actors and policy venues. Needless to say, power is not equally distributed among actors, and some sites of decision-making are more important than others. National governments still wield tremendous leverage both on the territories they govern and as the only legally authorized participants in international deliberations. Here again, the disparities are enormous but in all cases, national sovereignty is no longer absolute. Multilateral bodies, transnational corporations, and international treaties powerfully constrain the role of every nation state. Global governance is increasingly referred to as a multi-stakeholder process. The WSIS experience has transformed this framework most notably by sanctifying the place of global civil society as an organized force in this process.[6]

However, we caution against any hasty idealization of civil society's participation at the Summit. This is, after all, just a first experience that needs to be understood and analyzed. At the same time, it does shed light on emerging trends and points the way for events to come.

There can be no question that the creation of an autonomous, open and inclusive structure, the WSIS Civil Society Plenary, and its production of the Civil Society Declaration—despite their shortcomings—provide a model for the blending of issues and process which should inspire all those who are thinking about possibilities for a new global politics, not only in communication but in global affairs.[7]

As the first stone to be laid in the new edifice, the Geneva phase of the WSIS set the tone and triggered further discussions about global governance during the second phase in Tunis in 2005. Civil society sought to maximize its gains in Tunis with respect to issues that extend beyond the framework of the Summit itself, notably in the areas of North-South solidarity, Internet governance and human rights. The WSIS thus continued to be a laboratory of global governance and the actors certainly remained well aware of that fact.

One Summit, Two Worlds

One of the things that make the World Summit on the Information Society so interesting is that it lies at the intersection of two conflicting visions of communication whose confrontation rekindles debates that took place thirty years ago. Freed from the polarizations of the Cold War, the debates surrounding the New World Information and Communication Order (NWICO) have been regenerated with a new vigour at the WSIS.

The NWICO, which embroiled UNESCO in a period of acrimonious debates from the mid-1970s to the late 1980s, had been relegated until recently to the background at the United Nations.

Originally spearheaded by the Non-Aligned Movement, the creation of a New World Information and Communication Order was strongly supported by the developing countries, which found themselves in the majority at the United Nations following the waves of decolonization of the 1960s. By objecting to what they saw as economic and cultural colonization by the Western media corporations, and by challenging their state of dependence on the foreign media and press agencies, the developing countries opened the door to a questioning of the dominant paradigm of global communication.

In 1977, UNESCO responded to the polarization of the debates and the prospect of a major showdown by setting up an International Commission for the Study of Communication Problems, better known as the MacBride Commission after its Chairman, the Irish jurist Sean MacBride. When the Commission presented its report to the General Assembly of UNESCO in 1980, it submitted a coherent and well-argued case in favour of the demands of the developing countries and the advent of a New World Information and Communication Order.[8]

The rest is history. The Western countries, and their powerful media and industrial lobbies, publicly denounced the MacBride Report as containing the seeds of State authoritarianism. The United States withdrew from UNESCO in 1984, followed a year later by Great Britain and Singapore.

The NWICO debate was gradually eclipsed by consolidation of the Western capitalist conception of an international media order. The explosion of globalization, deregulation and the gradual rise of neoliberalism fed a utopian and mercantile vision of communication and information in which ICTs would meld the world's nations into one great information society, generating wealth and goods for all and spreading its benefits throughout all sectors.

This is basically the ideology that holds sway at the International Telecommunication Union today. The ITU denies civil society any access to its official bodies on the one hand, yet develops its policies through "public-private partnerships" and grants individual accreditations to business enterprises (provided they can afford to pay the dues), thus giving them the right to vote at the organization's meetings (although not on matters relating to the ITU's statutes and conventions).[9]

A number of civil society actors criticized the preponderance of the ITU over UNESCO in the organization of the WSIS out of concern over its possible influence. Their fears turned out to be justified to a large extent. Infrastructure development took precedence over culture, education, poverty and knowledge. The WSIS itself violated a rule of the UN Economic and Social Council by accrediting individual business entities, something which had never been done before at a United Nations Summit. Civil society, meanwhile, had to work tirelessly to get questions such as copyright, media concentration and cultural exclusion (subjects generally left to UNESCO) on the WSIS agenda.

So it was ultimately through the positions defended by civil society that questions left unresolved when the NWICO debate unravelled resurfaced at this high-level international event. The Civil Society Declaration *Shaping Information Societies for Human Needs* rejected the official terminology and its biases and revitalized a conception of information and communication that had seemed long dead and buried.

The concept of "information and communication societies", as an alternative and more encompassing notion than "information society", is central for civil society. It constitutes the point of semantic divide with respect to the official *Declaration of Principles*, it crystallizes the rejection of the intergovernmental vision, with its limitations and biases, and it invites us to move on to address more fundamental issues from which a new order and new solutions can flow:

> There is no single information, communication or knowledge society: there are, at the local, national and global levels, possible future societies; moreover, considering communication is a critical aspect of any information society, we use in this document the phrase "information and communication societies."[10]

By challenging the approach articulated in the official documents— everyone united in a single information society—, this footnote to the Civil Society Declaration expresses and consecrates the split between two visions of social life and the organization of human relations. It illustrates how the WSIS became the meeting point of two different social projects—one characterized by a singular information society, the other by the plural, participatory notion of information and communication societies. Where the government documents are above all about a managerial agenda, civil society envisions a possible future that defines a *"way of living together, an aspiration, a common project"*.

The *Civil Society Declaration* thus presents the vital human process of communication as the very foundation of a people-centred social order:

> At the heart of our vision of information and communications societies is the human being. The dignity and rights of all peoples and each person must be promoted, respected, protected and affirmed. Redressing the inexcusable gulf between levels of development and between opulence and extreme poverty must therefore be our prime concern.
>
> We are committed to building information and communication societies that are people-centred, inclusive and equitable. Societies in which everyone can freely create, access, utilise, share and disseminate information and knowledge, so that individuals, communities and peoples are empowered to improve their quality of life and to achieve their full potential. Societies founded on the principles of social, political, and

economic justice, and peoples' full participation and empowerment, and thus societies that truly address the key development challenges facing the world today. Societies that pursue the objectives of sustainable development, democracy, and gender equality, for the attainment of a more peaceful, just, egalitarian and thus sustainable world, premised on the principles enshrined in the Charter of the United Nations and in the Universal Declaration of Human Rights.

We aspire to build information and communication societies where development is framed by fundamental human rights and oriented to achieving a more equitable distribution of resources, leading to the elimination of poverty in a way that is not-exploitative and environmentally sustainable. To this end we believe technologies can be engaged as fundamental means, rather than becoming ends in themselves, thus recognising that bridging the digital divide is only one step on the road to achieving development for all. We recognise the tremendous potential of information and communications technologies (ICTs) in overcoming the devastation of famine, natural catastrophes, new pandemics such as HIV/AIDS, as well as the proliferation of arms.

We reaffirm that communication is a fundamental social process, a basic human need and a foundation of all social organization. Everyone, everywhere, at any time should have the opportunity to participate in communication processes and no one should be excluded from their benefits. This implies that every person must have access to the means of communication and must be able to exercise their right to freedom of opinion and expression, which includes the right to hold opinions and to seek, receive and impart information and ideas through any media and regardless of frontiers. Similarly, the right to privacy, the right to access public information and the public domain of knowledge, and many other universal human rights of specific relevance to information and communication processes, must also be upheld. Together with access, all these communication rights and freedoms must be actively guaranteed for all in clearly written national laws and enforced with adequate technical requirements.

Unlike the debate on the NWICO, which essentially took place between States in the total absence of civil society, the debate surrounding WSIS therefore had a completely different quality precisely because of the appearance of this new actor on the world stage. But the spectre of the NWICO continued to weigh on peoples' minds, even within civil society, where the advocates of a social conception of communication were pitted against the champions of unconstrained information abundance. And so a rich and urgent debate continues.

To the extent that the broader concept of "communication rights" is being developed and encompasses new issues (intellectual property rights and Internet governance, in particular), the WSIS has brought this debate up to date, in a new venue and involving new players.

The World Summit on the Information Society was therefore and above all a locus of confrontation between opposing paradigms of both communication and global governance, in a new political environment in which the involvement of civil society is expanding. Despite the obstacles and disappointments that we have chronicled in these pages, the bottom line is encouraging for the definition and emergence of new spaces in which people are able to participate more than ever before in taking charge of their own lives.

• APPENDIX •

"Shaping Information Societies for Human Needs"

We, women and men from different continents, cultural backgrounds, perspectives, experience and expertise, acting as members of different constituencies of an emerging global civil society, considering civil society participation as fundamental to the first ever held UN Summit on information and communication issues, the World Summit on the Information Society, have been working for two years inside the process, devoting our efforts to shaping a people-centred, inclusive and equitable concept of information and communication societies.[1]

Working together both on-line and off-line as civil society entities, practising an inclusive and participatory use of information and communication technologies, has allowed us to share views and shape common positions, and to collectively develop a vision of information and communication societies.

At this step of the process, the first phase of the Summit, Geneva, December 2003, our voices and the general interest we collectively expressed are not adequately reflected in the Summit documents. We propose this document as part of the official outcomes of the Summit. Convinced that this vision can become reality through the actions and lives of women and men, communities and people, we hereby present our own vision to all, as an invitation to participate in this ongoing dialogue and to join forces in shaping our common future.

---

[1] There is no single information, communication or knowledge society: there are, at the local, national and global levels, possible future societies; moreover, considering communication is a critical aspect of any information society, we use in this document the phrase "information and communication societies." For consistency with previous WSIS language, we retain the use of the phrase "Information Society" when directly referencing WSIS.

1. A Visionary Society

At the heart of our vision of information and communications societies is the human being. The dignity and rights of all peoples and each person must be promoted, respected, protected and affirmed. Redressing the inexcusable gulf between levels of development and between opulence and extreme poverty must therefore be our prime concern.

We are committed to building information and communication societies that are people-centred, inclusive and equitable. Societies in which everyone can freely create, access, utilise, share and disseminate information and knowledge, so that individuals, communities and peoples are empowered to improve their quality of life and to achieve their full potential. Societies founded on the principles of social, political, and economic justice, and peoples' full participation and empowerment, and thus societies that truly address the key development challenges facing the world today. Societies that pursue the objectives of sustainable development, democracy, and gender equality, for the attainment of a more peaceful, just, egalitarian and thus sustainable world, premised on the principles enshrined in the Charter of the United Nations and in the Universal Declaration of Human Rights.

We aspire to build information and communication societies where development is framed by fundamental human rights and oriented to achieving a more equitable distribution of resources, leading to the elimination of poverty in a way that is non-exploitative and environmentally sustainable. To this end we believe technologies can be engaged as fundamental means, rather than becoming ends in themselves, thus recognising that bridging the Digital Divide is only one step on the road to achieving development for all. We recognise the tremendous potential of information and communications technologies (ICTs) in overcoming the devastation of famine, natural catastrophes, new pandemics such as HIV/AIDS, as well as the proliferation of arms.

We reaffirm that communication is a fundamental social process, a basic human need and a foundation of all social organisations. Everyone, everywhere, at any time should have the opportunity to participate in communication processes and no one should be excluded from their benefits. This implies that every person must have access to the means of communication and must be able to exercise their right to freedom of opinion and expression, which includes the right to hold opinions and to seek, receive and impart information and ideas through any media and regardless of frontiers. Similarly, the right to privacy, the right to access public information and the public domain of knowledge, and many other universal human rights of specific relevance to information and communication processes, must also be upheld. Together with access, all these communication rights and freedoms must be actively guaranteed for all in clearly written national laws and enforced with adequate technical requirements.

Building such societies implies involving individuals in their capacity as citizens, as well as their organisations and communities, as participants and decision-makers in shaping frameworks, policies and governing mechanisms. This means creating an enabling environment for the engagement and commitment of all generations, both women and men, and ensuring the involvement of diverse social and linguistic groups, cultures and peoples, rural and urban populations without exclusion. In addition, governments should maintain and promote public services where required by citizens and establish accountability to citizens as a pillar of public policy, in order to ensure that models of information and communication societies are open to continuing correction and improvement.

We recognise that no technology is neutral with respect to its social impacts and, therefore, the possibility of having so-called "technology-neutral" decision-making processes is a fallacy. It is critical to make careful social and technical choices concerning the introduction of new technologies from the inception of their design through to their deployment and operational phases. Negative social and technical impacts of information and communications systems that are discovered late in the design process are usually extremely difficult to correct and, therefore, can cause lasting harm. We envision an information and communication society in which technologies are designed in a participatory manner with and by their end-users so as to prevent or minimise their negative impacts.

We envision societies where human knowledge, creativity, cooperation and solidarity are considered core elements; where not only individual creativity, but also collective innovation, based on cooperative work are promoted. Societies where knowledge, information and communication resources are recognised and protected as the common heritage of humankind; societies that guarantee and foster cultural and linguistic diversity and intercultural dialogue, in environments that are free from discrimination, violence and hatred.

We are conscious that information, knowledge and the means of communication are available on a magnitude that humankind has never dreamt of in the past; but we are also aware that exclusion from access to the means of communication, from information and from the skills that are needed to participate in the public sphere, is still a major constraint, especially in developing countries. At the same time information and knowledge are increasingly being transformed into private resources which can be controlled, sold and bought, as if they were simple commodities and not the founding elements of social organisation and development. Thus, as one of the main challenges of information and communication societies, we recognise the urgency of seeking solutions to these contradictions.

We are convinced that with the sufficient political will to mobilise this wealth of human knowledge and the appropriate resources, humanity could certainly achieve the goals of the Millennium Declaration, and even surpass them. As civil society organisations, we accept our part of responsibility in making this goal and our vision a reality.

2. Core Principles and Challenges

In accordance with this vision, it is essential that the development of information and communication societies be grounded in core principles that reflect a full awareness of the challenges to be met and the responsibility of different stakeholders. This includes the full recognition of the need to address gender concerns and to make a fundamental commitment to gender equality, non-discrimination and women's empowerment, and recognise these as non-negotiable and essential prerequisites to an equitable and people-centred development within information and communication societies. Such a commitment means consciously redressing the effects of the intersection of unequal power relations in the social, economic and political spheres, which manifests in differential access, choice, opportunity, participation, status and control over resources between women and men as well as communities in terms of class, ethnicity, age, religion, race, geographical location and development status.

We have identified the following as key areas of concern. We recognise and uphold the following principles; and we have identified certain priority areas for action by the international community.

2.1 Social Justice and People-Centred Sustainable Development

Within a social justice framework, human development implies cultural, social, economic, political and environmental living conditions that fulfill and empower individuals and communities. Despite the enormous advancements in knowledge and technology achieved by humanity, a majority of people continue to live in appalling conditions.

Social justice in the information and communication societies can only be pursued by taking into account geo-political and historical injustices along economic, social, political and cultural lines. Current global dynamics are characterised by tensions resulting from the inter-linkages of global economic liberalisation, cultural globalisation, increased militarism, rising fundamentalisms, racism and the suspension and violation of basic human rights.

The unequal distribution of ICTs and the lack of information access for a large majority of the world's population, often referred to as the digital divide, is in fact a mapping of new asymmetries onto the existing grid of social divides. These include the divide between the North and South, rich and poor, men and women, urban and rural populations, those with access to information and those without. Such disparities are found not only between different cultures, but also within national borders. The international community must exercise its collective power to ensure action on the part of individual states in order to bridge domestic digital divides.

Redressing all forms of discrimination, exclusion and isolation that different marginalised and vulnerable groups and communities experience will require more than the deployment of technology alone. Their full participation in information and communication societies requires us to reject at a fundamental level, the solely profit-motivated and market-propelled promotion of ICTs for development. Conscious and purposeful actions need to be taken in order to ensure that new ICTs are not deployed to further perpetuate existing negative trends of economic globalisation and market monopolisation. Instead, ICT development and applications should be oriented to advance the social, economic and cultural progress of the world's peoples and contribute to transforming the development paradigm.

Technological decisions should be taken with the goal of meeting the life-critical needs of people, not with the goal of enriching companies or enabling undemocratic control by governments. Therefore, fundamental decisions concerning the design and use of technologies must be made in cooperation with Civil Society, including individual end-users, engineers, and scientists. In particular, where community-based technologies are concerned the study and practice of community informatics must be applied in order to respond adequately to the particular characteristics and needs of communities in design processes.

2.1.1 Poverty Eradication

Poverty eradication must be a key priority on the WSIS agenda. Without challenging existing inequalities, no sustainable development embracing the new ICTs can be achieved. People living in extreme poverty must be enabled to contribute their experiences and knowledge in a dialogue involving all parties. Challenging poverty requires more than setting "development agendas". It requires a fundamental commitment to examine the current frameworks, to improve local access to information that is of relevance for the specific context, to improve training in ICT-related skills, and to allocate significant financial and other resources. Also, because volunteers are working at the grassroots level, they play an important role in social inclusion.

Financial resources, linked with social and digital solidarity, need to be channelled through existing and new financial mechanisms that are managed transparently and inclusively by all sectors of society. Among the frameworks that need to be examined in terms of their potentially adverse effects on equitable development are the current arrangements for recognition and governance of monopolised knowledge and information, including the work of WIPO and the functioning of the TRIPS agreement.

2.1.2 Global Citizenship

Information and communication societies have the potential to catalyse and help release the enormous financial, technical, human and moral resources required for sustainable development. These resources will only be freed up as the peoples of the world develop a profound sense of responsibility for the fate of the planet and the well-being of the entire human family. In this regard, there is a need for the development in the individual and in communities, as well as governments, of a global consciousness, and a sense of world citizenship. Since the body of humankind is one and indivisible, each member of the human race is born into the world as a trust of the whole and is best served by ensuring the equal importance of each member through the proactive exercise and application of international human rights standards.

2.1.3 Gender Justice

Equitable, open and inclusive information and communication societies must be based on gender justice and be particularly guided by the interpretation of principles of gender equality, non-discrimination and women's empowerment as contained in the Beijing Declaration and Platform for Action (Fifth World Conference on Women) and the Convention on the Elimination of All Forms of Discrimination Against Women (CEDAW). Actions must demonstrate not only a strong commitment but also a high level of consciousness to an intersectional approach to redressing discrimination resulting from unequal power relations at all levels of society. Proactive policies and programmes across all sectors must be developed for women as active and primary agents of change in owning, designing, using and adapting ICT. To empower girls and women throughout their life cycle, as shapers and leaders of society, gender responsive educational programs and appropriate learning environments need to be promoted. Gender analysis and the development of both quantitative and qualitative indicators in measuring gender equality through an extensive and integrated national system of monitoring and evaluation are "musts."

2.1.4 Importance of Youth

We recognise also that young people are the future workforce and leading creators and earliest adopters of ICTs. They must therefore be empowered as learners, developers, contributors, entrepreneurs and decision-makers. We must focus especially on young people who have not yet been able to benefit fully from the opportunities provided by information and communication societies. In particular, we must seek to assist and empower youth from disadvantaged backgrounds, especially young people in developing countries. Equality of opportunity for girls and young women must be integral to our efforts, and we must create a greater awareness of their specific needs and potential in the field of ICT. Issues facing young workers in ICT industries, such as low pay, poor working conditions, and a lack of job stability and collective representation, must also be addressed. As main users of ICTs, young people are most affected and vulnerable to the health risks exposed by their use. Therefore we commit to develop and use only those ICTs that ensure the well-being, protection, and harmonious development of all children.

2.1.5 Access to Information and the Means of Communication

Access to information and the means of communication as a public and global commons should be participatory, universal, inclusive and democratic. Inequalities in access must be addressed in terms of the North/South divide as well as in terms of enduring inequalities within developed and less developed nations. Barriers that need to be overcome are of an economic, educational, technical, political, social, ethnic, and age nature, and inequitable gender relations are embedded into all of these and need to be specifically addressed.

Universal access to information that is essential for human development must be ensured. Infrastructure and the most appropriate forms of information and communication technologies must be accessible for all in their different social context, and the social appropriation of these technologies must be encouraged. This implies addressing diverse realities experienced by distinct social groups such as indigenous peoples, diasporas and migrants, and privileging local or targeted solutions. Traditional media and community-based information and communication initiatives have a vital role to play in these respects, and so does the effective use of the new ICTs. The regulatory and legal framework in all information and communication societies must be strengthened to support broad-based sharing of technologies, information, and knowledge, and to foster community control, respectful of human rights and freedoms.

Specific needs and requirements of all stakeholders, including those with disabilities, must be considered in ICT development. Accessibility and inclusiveness of ICTs is best done at an early stage of design, development and production, so that the information and communication society becomes the society for all, at minimum cost.

The need to access, send and receive information represents a particularly vital challenge to vulnerable people such as refugees, those displaced by war, and asylum seekers who often do not know their rights, which are frequently violated. Access to means of communication for these groups is necessary for the defence and promotion of their rights, in order to make legitimate claims in conformity with international law.

2.1.6 Access to Health Information

The delivery of life-critical mental and physical health information can be facilitated and improved through ICT-based solutions. Lack of access to information and communication has been identified as a critical factor in the public mental and physical health crises around the world. Experts have suggested that providing citizens of developing countries with community level points of access to mental and physical health information would be a critical starting point for addressing the mental and physical health care crises. However, such access points should support more than one-way flows of information (for example, from expert to community or patient). Communities must be allowed to participate in the selection and creation of communication flows that they find useful and necessary to address the prevention, treatment, and promotion of mental and physical health care for all people. Open access to medical information is absolutely essential so that all known data are available to medical doctors and practitioners.

2.1.7 Basic Literacy

Literacy and free universal access to education is a key principle. Knowledge societies require an informed and educated citizenry. Capacity-building needs to include skills to use ICTs, media and information literacy, and the skills needed for active citizenship including the ability to find, appraise, use and create information and technology. Approaches that are local, horizontal, gender-responsive and socially driven and mediated should be prioritised. A combination of traditional and new media as well as open access to knowledge and information should be encouraged. Libraries—both real and virtual—have an important role to play to ensure access to knowledge and information available to everyone. At the international and multilateral level, the public domain of knowledge and culture needs to be protected. People-centred information technologies can foster eradication of illnesses and epidemics, can help give everyone food, shelter, freedom and peace.

Literacy, education and research are fundamental components of information, communication and knowledge societies. Knowledge creation and acquisition should be nurtured as a participatory and collective process and not considered a one-way flow or confined to one section of capacity building. Education (formal, informal, and lifelong) builds democracy both by creating a literate citizenry and a skilled workforce. But only an informed and educated citizenry with access to the means and outputs of pluralistic research can fully participate in and effectively contribute to knowledge societies.

Urgent attention should be paid to the potential positive and negative impacts of ICTs on the issues of illiteracy in regional, national and international languages of the great majority of the world's peoples. Literacy, education, and research efforts in the information and communication societies must include a focus on the needs of people who have physical impairments and all means of transcending those impairments (for example, voice recognition, e-learning, and open university training) must be promoted.

2.1.8 Development of Sustainable and Community-based ICT Solutions

In order that communities and individuals may fully enjoy the benefits of the information and communication society, ICTs must be designed and manufactured according to environmentally sustainable principles. Technological solutions must also be sustainable in the sense that communities are able to support their use and evolution.

Equipment recycling must meet environmental standards. The production of technologies must not consume an unsustainable amount of energy or natural resources.

It is essential to develop concrete proposals and policies to improve resource efficiency and develop renewable energy resources. This involves "dematerialising" (for example, using less paper) and reducing ICT-related waste; increasing the useful life of hardware; improving recycling conditions; ensuring safe disposal of discarded ICT hardware and parts; and encouraging the development of alternatives to toxic ICT components. This also implies giving the highest priority to creating and using renewable energy resources to address the basic needs of populations living in developing countries. Renewable energy resources should be used for ICT-based dissemination of information and communications, including radio and television. Africa can particularly benefit from solar power due to its high level of exposure to direct solar radiation. By mobilising regional synergies, complemented by the necessary technical and financial cooperation, Africa could play a leading role in this strategic domain in the next decade.

Communities must have the ability to participate directly in the development and maintenance of ICT-based solutions to their own problems. In order that communities may create and sustain their own solutions using ICTs, they must be empowered to develop their own productive forces and control the means of production within information societies. This must include the right to participate fully in the development and sustenance of ICT-based projects through democratic processes, including decision making with respect to economic, cultural, environmental, and other issues. ICTs should be used as an instrument for the creation of genuine and sustainable sources of work, thus providing new labour opportunities.

In order that communities and individuals may create economically and technically sustainable solutions, they must have the right to use Free Software. This makes software more affordable, and, allows people to participate in its development and maintenance[2]. ICT-based innovation should adhere to the use of international technical standards for hardware, software, and processes, which are open, freely implementable, publicly documented, interoperable, non-discriminatory and demand-driven.

It is important to support community-based communications using both traditional and new media and communication technologies. There is a need for the development and nurturing of the discipline of community informatics, which focuses on the particular characteristics and needs of communities,

---

[2] In this document, we use the term "Free Software" to refer to the specific concept defined by the Free Software Foundation. Free Software is software that is licensed in such a way that people have the freedom to run, copy, distribute, study, change and improve it. Free Software implies access to source code as does "open source software"; however, open source software as the term is popularly used is not necessarily Free Software in our definition. Some organisations release open source software without permitting all of these actions. See http://www.fsf.org and http://www.fsfeurope.org for in-depth discussions of this concept.

in relation to design, development, deployment, and operation of ICTs, as well as local content production.

2.1.9 Conflict Situations

We recognise that the use of media can be both positive and negative in conflict situations, including post-conflict peace building. We therefore insist that the rights of journalists and of all people to gather and communicate information, using any media, be especially respected during conflicts. These rights should be inviolate at all times but are crucial during war, violent conflict, and non-violent protest.

We are particularly concerned about the deployment of "information warfare" technologies and techniques, including the purposeful jamming, blocking, or destruction of civilian communication systems during conflict situations; the use of "embedded" journalists coupled with the targeting of non-embedded journalists; the use of media and communication systems to promote hatred and genocide; by military, police, or other security forces, be they governmental, privately owned, or non-state actors, during conflict situations both international and domestic.

Information intervention in conflict situations should be bound by international law, and the WSIS should encourage work on a future convention against information warfare to address these concerns. At the same time, the WSIS should not only limit information warfare and the control of media in conflict situations, but also actively promote media and communications for peace. To that end, we encourage governments to decrease public subsidy for military communications technology, and instead spend money directly on developing peaceful communications tools and applications.

2.2 Centrality of Human Rights

An information and communication society should be based on human rights and human dignity. With the Charter of the United Nations and the Universal Declaration of Human Rights as its foundation, it must embody the universality, indivisibility, interrelation and interdependence of all human rights—civil, political, economic, social and cultural—including the right to development and linguistic rights. This implies the full integration, concrete application and enforcement of all rights and the recognition of their centrality to democracy and sustainable development. Information and communication societies must be inclusive, so that all people, without distinction of any kind, can achieve their full potential. The principles of non-discrimination and diversity must be mainstreamed in all ICT regulation, policies, and programmes.

2.2.1 Freedom of Expression

Article 19 of the Universal Declaration of Human Rights is of fundamental and specific importance, since it forms an essential condition for human rights-based information and communication societies. Article 19 requires that everyone has the right to freedom of opinion and expression and the right to seek, receive and impart information and ideas, through any media and regardless of frontiers. This implies free circulation of ideas, pluralism of the sources of information and the media, press freedom, and availability of the tools to access information and share knowledge. Freedom of expression on the Internet must be protected by the rule of law rather than through self-regulation and codes of conduct. There must be no prior censorship, arbitrary control of, or constraints on, participants in the communication process or on the content, transmission and dissemination of information. Pluralism of the sources of information and the media must be safeguarded and promoted.

2.2.2 Right to Privacy

The right to privacy, enshrined in Article 12 of the Universal Declaration of Human Rights, is essential for self-determined human development in regard to civic, political, social, economic and cultural activities. The right to privacy faces new challenges in information and communication societies, and must be protected in public spaces, online, offline, at home and in the workplace. Every person must have the right to decide freely whether and in what manner he or she wants to receive information and communicate with others. The possibility of communicating anonymously must be ensured for everyone. The power of the private sector and of governments over personal data increases the risk of abuse, including monitoring and surveillance. Such activities must be kept to a legally legitimised minimum in a democratic society, and must remain accountable. The collection, retention, processing, use and disclosure of personal data, no matter by whom, should remain under the control of and determined by the individual concerned.

2.2.3 Right to Participate in Public Affairs

Good government administration and justice in a democratic society implies openness, transparency, accountability, participation and compliance with the rule of law. Respect for these principles is needed to enforce the right to take part in the conduct of public affairs. Public access to information produced or maintained by governments should be enforced, ensuring that the information is timely, complete and accessible in a format and language the public can understand. This further applies to access to documents of corporations relating to their activities affecting the public interest, especially in situations where the government has not made such information public.

2.2.4 Workers' Rights

ICTs are progressively changing our way of working. The creation of fair, secure, safe and healthy working conditions, in the manufacture of equipment and software, and in the utilisation of ICTs in the workplace in general, which respect international labour standards, for instance through tripartite social dialogue, is fundamental. ICTs should be used to promote awareness of, respect for and enforcement of human rights standards and international labour standards. Human rights, such as privacy, freedom of expression, linguistic rights, the right for on-line workers to form and join trade unions and the right of trade unions to function freely, including communicating with employees, must be respected in the workplace.

2.2.5 Rights of Indigenous Peoples

The evolution of information and communication societies must be founded on the respect and promotion of the recognition of the Rights of Indigenous Peoples and their distinctiveness as outlined in international conventions. Indigenous Peoples have fundamental rights to protect, preserve and strengthen their own language, culture and identity. ICTs should be used to support and promote diversity and the rights and means of Indigenous Peoples to benefit fully and with priority from their cultural, intellectual and so-called natural resources.

2.2.6 Women's Rights

In order to realise women's rights in the information and communication societies, as spelled out in the Convention on the Elimination of All Forms of Discrimination Against Women (CEDAW) and the Beijing Declaration and Platform for Action (Fifth World Conference on Women), it is crucial to acknowledge and address the differences, disparities and disadvantages that women experience. This means taking into account the ways in which women are different from men, and how these differences translate into differential levels of access, opportunity, participation and use of ICTs. It must be ensured that policy or legal interventions and programmes consciously address these differences. To ensure effective equality of women, and thereby enabling women's full ability to claim and exercise their human rights, it is necessary to adopt a substantive equality approach in the analysis, which informs the content of ICT policy and programmes. This approach implies that actions to promote women's rights must transform the unequal power relation between women and men. Women need not only equality of opportunity, but also equality of access to opportunities and the ability to fully participate in availing such opportunities.

2.2.7 Rights of the Child

Information and communication societies must respect and promote the principles of the Convention on the Rights of the Child. Every child is entitled to a happy childhood and to enjoy the rights and freedoms available to all persons under the Universal Declaration of Human Rights. All persons, civil society, private sector and governments should commit to uphold the Rights of the Child in information and communication societies.

2.2.8 Rights of Persons with Disabilities

In inclusive information and communication societies, the rights of persons with disabilities to have full and equal access to information and communications including ICTs, regardless of types and degree of disabilities, must be ensured by public policies, laws and regulations at all levels. In order to achieve this goal, a Universal Design principle and the use of assistive technologies must be seriously promoted and supported throughout the whole process of building and nurturing information and communication societies in which persons with disabilities and their organisations must be allowed to participate fully and on equal terms with non-disabled people.

2.2.9 Regulation and the Rule of Law

National regulation should be in full compliance with international human rights standards, adhering to the rule of law. Information and communication societies must not result in any discrimination or deprivation of human rights resulting from the acts or omission of governments or of non-state actors under their jurisdictions. Any restriction on the use of ICTs must pursue a legitimate aim under international law, be prescribed by law, be strictly proportionate to such an aim, and be necessary in a democratic society.

2.3 Culture, Knowledge and the Public Domain

Information and communication societies are enriched by their diversity of cultures and languages, retained and passed on through oral tradition or recorded and transmitted through a variety of media, and together contributing to the sum of human knowledge. Human knowledge is the heritage of all humankind and the reservoir from which all new knowledge is created. The preservation of cultural and linguistic diversity, the freedom of the media and the defence and extension of the public domain of global knowledge are as essential, for information and communication societies, as the diversity of our natural environment.

2.3.1 Cultural and Linguistic Diversity

Cultural and linguistic diversity is an essential dimension of people-centred information and communication societies. Every culture has dignity and value that must be respected and preserved. Cultural and linguistic diversity is based, among other things, on the freedom of information and expression and the right of everyone to freely participate in the cultural life of the community, at local, national and international levels. This participation includes activities both as users and producers of cultural content. ICTs including traditional communications media have a particularly important role to play in sustaining and developing the world's cultures and languages.

2.3.1.1 Capacity Building and Education

Cultural and linguistic diversity should not only be preserved; it needs to be fostered. This implies capacity to express oneself, in one's own language, at any time, by any means, including traditional media and new ICTs. In order to become a contributor and a creator in the information and communication societies, not only technical skills are needed, but critical and creative competence. Media education in the sense of the UNESCO Grunwald Declaration must be given specific attention in education and training programs. Cultural and linguistic diversity also implies equal access to the means of expression and of dissemination of cultural goods and services. Priority should be given to community-driven initiatives.

2.3.1.2 Language

Plurality of languages is at the core of vibrant information and communication societies. ICTs can be applied to bridge cultural and linguistic divides, given the right priorities. In the past, ICT development has too often reinforced inequalities, such as dominance of roman letter based languages (especially English) and marginalization of local, regional and minority languages. Priority should be given in ICT research and development to overcoming barriers and addressing inequalities between languages and cultures.

2.3.1.3 International Law and Regulation

International law and regulation should strengthen cultural, linguistic and media diversity, in accordance with existing international declarations and covenants, in particular Article 19 and Article 27 of the Universal Declaration of Human Rights; Articles 19 and 27 of the International Covenant on Civil and Political Rights; Articles 13 and 15 of the International Covenant on Economic, Social and Cultural Rights; and Articles 5 and 6 of the Universal Declaration of Cultural Diversity adopted by UNESCO in 2001. International trade agreements should treat culture, including audio-visual content and services, not simply as a commodity, but should take account of the need for cultural, linguistic and media diversity. The establishment of an International Convention on Cultural Diversity should be accelerated, with a view to achieving an effective and binding international agreement. Existing international copyright regulation instruments including TRIPS and WIPO should be reviewed to ensure that they promote cultural, linguistic and media diversity and contribute to the development of human knowledge.

2.3.2 Media

2.3.2.1 The Role of the Media

Freedom of Expression and Freedom of the Media are central to any conception of information and communication societies. The media is an integral enabling mechanism for a global communications vision. Its role in producing, gathering and distributing diverse content in which all citizens are included and can actively participate, is vital. Especially for the developing countries, broadcast radio and television will continue to be the most effective ways to deliver high-quality information. All forms of media can make crucial contributions to social cohesion and development in the digital era.

• Appendix • 151

Article 19 is the foundation for five regional declarations on media freedom and plurality that must continue to frame the role of the media in all its means of delivery. These texts[3] have been unanimously endorsed by the member states of UNESCO.

Security and other considerations should not be allowed to compromise freedom of expression and media freedom. Media pluralism and diversity should be guaranteed through appropriate laws to avoid excessive media concentration.

Editorial independence of media professionals and creators must be protected and the formulation of professional and ethical standards in journalism and other media production must be the responsibility of media workers themselves. Online authors, journalists and editors should have the same contractual rights and social protections as other media workers.

Public service broadcasting has a specific and crucial role to play in ensuring the participation of all in the information and communication societies. State-controlled media should be transformed into editorially independent public service organisations.

2.3.2.2 Community Media

Community media, that is media which are independent, community-driven and civil society-based, have a particular role to play in enabling access to and participation for all in information and communication societies, especially the poorest and most marginalized communities. Community media can be vital enablers of information, voice and capacities for dialogue. Legal and regulatory frameworks that protect and enhance community media are especially critical for ensuring vulnerable groups access to information and communication.

Governments should ensure that legal frameworks for community media are non-discriminatory and provide for equitable allocation of frequencies through transparent and accountable mechanisms. Targets should be established for the opening up of broadcast licenses to allow for the operation of community broadcasting where this is not currently permitted. Spectrum planning and regulation should ensure sufficient spectrum and channel capac-

---

[3] The Windhoek Declaration on the Promotion of Free and Pluralistic African Press, 1991; the Declaration of Alma Ata on Promoting Independent and Pluralistic Asian Media, 1992; the Declaration of Sana'a on Promoting Independent and Pluralistic Media, 1994; the Sofia Declaration on Promoting European Pluralistic and Independent Media, 1997 (adopted in 1995 and 1997).

ity, and appropriate technical standards, for community media to develop in both the analogue and the digital environment.

A Community Media Fund should be established through a donor civil society partnership to invest in and support community-driven media, information and communication initiatives using traditional media and new ICTs including projects that make provision for the poorest communities, for cultural and linguistic diversity and for the equal participation of women and girls. Community-based media and communication centres should be encouraged and assisted to combine traditional media technologies, including radio and television, with access to new ICTs.

2.3.3 The Public Domain of Global Knowledge

A rich public domain of knowledge available to all is essential to sustainable information societies, to bridge the digital divide and to provide the grounds for a positive development of intellectual creativity, technological innovation and effective use of that technology. In information societies, new digital forms of storing information mean that this can be copied and transmitted in innovative ways that challenge existing customs and laws. The increasing privatisation of knowledge production threatens to restrict the availability of research results. Attempts have been made to commercially exploit traditional indigenous knowledge without consulting the communities, who are the owners of that knowledge.

2.3.3.1 Indigenous Peoples' Knowledge

Indigenous peoples are the guardians of their traditional knowledge and have the right to protect and control that knowledge. Existing intellectual property regimes are insufficient for the protection of indigenous people's cultural and intellectual property rights.

Traditional knowledge should be protected from any attempt at patenting. Indigenous peoples should freely decide whether their heritage should become part of the public domain or not. They should decide whether or not it should be exploited commercially and in what way.

We should give particular attention to measures to maintain knowledge diversity and to protect the cultural, intellectual and so-called natural resources of indigenous peoples, especially botanical and agricultural knowledge, from commercial exploitation and appropriation.

We urge the United Nations to establish specific legal frameworks, in accordance with Article 26.4 of the Agenda 21 of the Earth Summit, to recognise indigenous peoples' rights to self-determination and ancestral territories, as a necessary prerequisite to ensure the protection, preservation and development of their traditional knowledge in information and communication societies.

2.3.3.2 Copyright, Patents and Trademarks

Limited intellectual monopolies, also known as intellectual property rights, are granted only for the benefit of society, most notably to encourage creativity and innovation. The benchmark against which they must be reviewed and adjusted regularly is how well they fulfill this purpose. Today, the vast majority of humankind has no access to the public domain of global knowledge, a situation that is contributing to the growth of inequality and exploitation of the poorest peoples and communities. Yet instead of extending and strengthening the global domain, recent developments are restricting information more and more to private hands. Patents are being extended to software (and even to ideas), with the consequent effect of limiting innovation and reinforcing monopolies. Drugs that could save millions of lives are denied to disease sufferers because pharmaceutical companies that hold the patents resist making them available to those countries that can not pay high prices. Copyright periods have been extended again and again, making them practically indefinite and defeating their original purpose.

2.3.3.3 Software

Software provides the medium and regulatory framework for digital information, and access to software determines who may participate. Equal access to software is fundamental for inclusive and empowering digital information and communication societies, and a diversity of platforms is essential to this.

We must recognise the political and regulatory impact of software on digital societies and build, through public policy and specific programs, awareness of the effects and benefits of different software models. In particular, Free Software, with its freedoms of use for any purpose, study, modification and redistribution should be promoted for its unique social, educational, scientific, political and economic benefits and opportunities. Its special advantages for developing countries, such as low cost, empowerment and the stimulation of sustainable local and regional economies, easier adaptation to local cultures and creation of local language versions, greater security, capacity building, etc, need to be recognised, publicised and taken advantage of. Governments should promote the use of Free Software in schools and higher education and in public administration.

The UN should carry out a fundamental review of the impact on poverty and human rights of current arrangements for recognition and governance of monopolised knowledge and information, including the work of WIPO and the functioning of the TRIPS agreement. Efforts should be made to ensure that limited intellectual monopolies stimulate innovation and reward initiative, rather than keeping knowledge in private hands until it is of little use to society.

2.3.3.4 Research

Increasing private sector participation in scientific research is leading to patents and scientific knowledge being held in private hands instead of being available in the public domain, and increasing competition among scientists and scientific teams sometimes results in poor scientific practices, secrecy and the patenting of discoveries that would previously have been available to all. Research should continue to be based on cooperation, openness and transparency.

Public bodies such as libraries, scientific research centres, universities, should be able to contribute to enrich the common good of culture and knowledge, by putting into the public domain the results of their publicly funded activities. The public domain of global knowledge should be defended and extended through public policy, awareness-building and investments in programmes. These should ensure that any work funded by public or philanthropic bodies enters the public domain and should increase accessibility of information in online and offline media by means of Free Documentation, public libraries and other initiatives to disseminate information, such as Open Access journals and Open Archives giving access to scientific and other public domain information. All scientific data, such as genomes of living beings, should be freely accessible to all in Open Access databases.

2.4 Enabling environment

2.4.1 Ethical Dimensions

Information and communication societies are about how our societies create, share and utilise the information, cultural production and knowledge, which in turn shape the evolution of those societies. The value-base of the information society must be founded on the principles contained in the ensemble of internationally agreed-upon conventions, declarations, and charters.

More specifically, equal, fair and open access to knowledge and information resources,—whatever the technical means used to store and transmit them—must be established as fundamental principles of such societies. Technological, financial and regulatory considerations must conform to these principles.

Transparent and accountable governance, ethical business and accounting practices in communications sector firms and ethical media practice are of particular relevance in this context. Codes of ethics and standards should be adopted in these cases and mechanisms should be established to monitor their application as well as appropriate sanctions for their violation. Formulation of ethics and standards in journalism and other media production should be the responsibility of media workers themselves.

Respect for diversity must be a central criterion in establishing the principles and mechanisms for resolving conflicts that arise in information societies. Such societies, if they are built on values such as cooperation, equity, honesty, integrity, respect and solidarity, can have a significant impact on the quality of interaction between cultures and the promotion of meaningful dialogue among civilisations, and thus contribute to bringing about world peace.

2.4.2 Democratic and Accountable Governance

National and international regulations for information and communication societies should be in full compliance with international human rights standards. Openness, transparency, accountability and the rule of law should be the guiding principles for the democratic governance of societies at all levels, from the local to the national and international. Inclusive, participatory and peaceful information and communication societies rest on the responsiveness of governing bodies as well as on the commitment of all actors involved in governance, both of governmental and non governmental nature, to progressively implement greater political, social and economic equity.

A democratic perspective on information and communication societies, in which information is crucial for citizens, is necessary in order to make choices grounded on the awareness of alternatives and opportunities. Information and communication are the foundation for transparency, debate and decision-making. They can contribute to a culture and a practice of cooperation, basis for a renewal of democracy. Information and communication technologies offer potential benefits to the world's communities that will only be exploited if there is a political will to do so.

In this spirit, the aim of WSIS "to develop a common vision and understanding of the Information Society", and the methods to achieve such a vision, requires shared communication values and mechanisms including the right to communicate, respect for freedom of opinion and expression in all of its dimensions, and a commitment to transparency, accountability, and democracy.

2.4.3 Infrastructure and Access

The dramatic lack of a reliable infrastructure is the main physical obstacle for ICT-based services to be offered to populations living in Africa. Here, the fragmented and incomplete structure and the unreliability of the existing infrastructure and access networks constitute the underlying structure of the so-called Digital Divide.

(Tele) communications infrastructure is essential for disseminating ICT-based services and is central in achieving the goal of universal, sustainable, ubiquitous and affordable access to and usage of these technologies and services by all. Furthermore, energy is a prerequisite for infrastructure and access.

Most voice, data and Internet traffic between African countries is currently routed outside of the continent because of the lack of an efficient African backbone network, increasing the cost of this traffic. Increased cost always limits access. Existing efforts to build an African network infrastructure must be supported and expanded (e.g. Internet exchange points).

The implementation and roll-out of (tele)communications infrastructure and access in DCs will require financial investments consistent with the huge needs in this area. In order to reduce the amount of financial resources needed, investments should be optimised by consolidating projects nationally or (sub) regionally, and by technological (re-) designing and updating. Furthermore, synergy between different sectors should be systematically exploited from the project phase, particular attention being paid to the energy and transport sectors that show very close links. Finally, the particularly strong synergy and technological similarity between ICT and Radio-TV networks should lead governments and planning authorities to deploy and use a common infrastructure for both their services to be transported and disseminated.

Community telecentres (public access centres) have become spaces for the effective access and strategic use of information and communication technologies with emphasis on the democratisation of communications. Governments should guarantee policies for the development of telecentres, among others, to provide equitable and affordable access to infrastructure and ICTs; to encourage digital inclusion policies for the population, independently of gender, ethnic aspects, language, culture and geographical situation. This would promote the discussion and active participation of communities in public policy processes related to the implementation and role of telecentres for local development.

Orbital satellite paths should be recognised as a public resource and should be allocated to benefit the public interest through transparent and accountable frameworks. Moreover, spectrum planning and regulation should ensure equitable access among a plurality of media including sufficient satellite capacity reserved for community media. A fixed percentage of orbital resources, satellite capacity and radio frequency spectrum should be reserved for educational, humanitarian, community and other non-commercial use.

The expansion of the global information infrastructure should be based on principles of equality and partnership and guided by rules of fair competition and regulation at both national and international levels.

The integration of access, infrastructure and training of the citizenry and the generation of local content, in a framework of social networks and clear public or private policies, is a key basis for the development of egalitarian and inclusive information societies.

2.4.4 Financing and Infrastructure

Existing and new financing measures should be envisaged and appraised. The "Digital Solidarity Fund" has been proposed by Africa. Such a fund could be a real hope for African peoples if it clearly states its goals, is transparently managed, and aims to foster primarily public services, especially for populations living in underserved and isolated areas. In addition, we stress the significant role that diaspora populations from all the world's regions can play in financing ICT programmes and projects.

In order to optimise scarce financial resources, appropriate cost-effective technological options should be used, while avoiding duplication of infrastructure. Additionally, synergies between different sectors and networks can be exploited to this end, with particular attention to the energy and transport sectors, given their close links with the telecommunications sector.

A Community Media Fund should be established through a donor civil society partnership to invest in and support community-driven and community-based media, and information and communication initiatives using both traditional media and new ITC's. Effort should be made to eliminate the duplication of infrastructures and to consolidate projects in a national or regional frame to encourage investment funding. Where possible, ICT and radio/TV networks should use common infrastructure for dissemination.

2.4.5 Human Development – Education and Training

Literacy, education and research are fundamental and interrelated components of the information exchanges necessary to build knowledge societies.

Knowledge creation and acquisition should be nurtured as a participatory and collective process; it should not be considered a one-way flow or confined to one section of capacity building. Education, in its different components—formal, informal, and lifelong—is fundamental to building democratic societies both by creating a literate citizenry and a skilled workforce.

To utilise the full potential of e-learning and long-distance education, they must be complemented by traditional educational resources and methods, in a local context of media pluralism and linguistic diversity.

Only informed and educated citizens with access to empowering education, a plurality of means of information, and the outputs of research efforts can fully participate in and effectively contribute to knowledge societies. Therefore it is also essential to recognise the right to education as stated both in the Declaration on the Right to Development and the Universal Declaration of Human Rights.

Capacity building initiatives designed to empower individuals and communities in the information society must include, in addition to basic literacy and ICT skills, media and information literacy, the ability to find, appraise, use and create information and technology. In particular, educators, students and researchers must be able to use and develop Free Software, which allows the unfettered ability to study, change, copy, distribute, and run software. Finally, capacity building initiatives should be designed to stimulate the desire for general learning and respond to specific as well as special needs: those of young and elderly people, of women, of people with impairments, of indigenous peoples, of migrant communities, of refugees and returnees in post-conflict situations, in a life-long perspective. Volunteers can help transmit knowledge and enhance capacity, in particular of marginalized groups not reached by government training institutions.

Capacity building in the information and communication societies requires people who are competent in teaching media and communication literacy. Therefore training of trainers and training of educators in every level is equal important in order to reach out to people at the limits of the information society.

Libraries are an important tool to fight digital divide and to ensure continuous, out-of-market-ruled access to information, by freeing the results of research funded by public support, by sharing content and educational materials to promote literacy, build capacities and bring autonomy to learners of all kinds, world wide. This also entails convincing content producers to be active participants in the open access paradigm of knowledge.

Global barriers to knowledge and education must be transparently evaluated by looking beyond technological obstacles at legal and institutional gridlocks (like intellectual property laws and international standards) and promoting a new balance of intellectual properties as a common ground for creators to protect their works and for civil society to benefit from their contributions.

Civil society sees the need for alternative models for the production and exchange of knowledge and information. To secure and finance the global knowledge commons, civil society actors support new open and self-organised publishing models in science and software production and community-based communications, with in-built maintenance programs and upgrading capacities.

2.4.6 Information Generation and Knowledge Development

Research must be promoted in all fields related to the information and communication societies, and its development must be sensitive to the social uses of ICTs. In particular, research on community informatics must be supported[4]. This would include the development of a research agenda among practitioners, scholars, and communities; the cataloguing of community informatics projects and identification of both factors for failure and success; and support for research projects and systems trials. Fundamental research should be strengthened by expanding open access to primary scientific data and publications. Public bodies such as libraries, scientific research centres, universities should foster independent investigation, build a pluralistic body of knowledge and promote the results of activities which have been funded by public money. This body of knowledge should be made available in all public spaces, or spaces with public access (community centres, universities, schools, museums, libraries, media centres, and other dedicated entities), through appropriate and plural modes of access, avoiding the risk of high dependency on digital technology alone.

2.4.7 Global Governance of ICT and Communications

International "rules of the game" play an increasingly central role in the global information economy. In recent years, governments have liberalised traditional international regulatory regimes for telecommunications, radio fre-

---

[4] Community informatics refers here to the interdisciplinary study and practice of the design, implementation, and management of information and communication technologies developed by communities to solve their own problems. This field takes into account social science research about the social impacts of ICTs—also known as social informatics—as well as information and communication systems analysis and design techniques.

quency spectrum, and satellite services, and have created new multilateral arrangements for international trade in services, intellectual property, "information security," and electronic commerce. At the same time, business groups have established a variety of "self-regulatory" arrangements concerning Internet identifiers (names and numbers), infrastructure, and content.

It is not acceptable for these and related global governance frameworks to be designed by and for small groups of powerful governments and companies and then exported to the world as *faits accomplis*. Instead, they must reflect the diverse views and interests of the international community as a whole. This overarching principle has both procedural and substantive dimensions.

Procedurally, decision-making processes must be based on such values as inclusive participation, transparency, and democratic accountability. In particular, institutional reforms are needed to facilitate the full and effective participation of marginalized stakeholders like developing and transitional countries, global civil society organisations, small and medium-sized enterprises, and individual users.

Substantively, global governance frameworks must promote a more equitable distribution of benefits across nations and social groups. To do so, they must strike a better balance between commercial considerations and other legitimate social objectives. For example, existing international arrangements should be reformed to promote: efficient management of network interconnections and traffic revenue distribution, subject to the mutual agreement of corresponding operators; equitable allocations of radio frequency spectrum and satellite orbital slots that fully support developmental and non-commercial applications; fair trade in electronic goods and services, taking into account the developing countries' need for special and differential treatment; an open public domain of information resources and ideas; and the protection of human rights, consumer safety, and personal privacy. In parallel, new diverse international arrangements are needed to promote: financial support for sustainable e-development, especially but not only in less affluent nations; linguistic, cultural, and informational diversity; and the curtailment of concentrated market power in ICT and mass media industries.

In light of the relevant controversies in the WSIS process, special attention must be given to improving the global coordination of the Internet's underlying resources. It must be remembered that the Internet is not a singular communications "platform" akin to a public telephone network; it is instead a highly distributed set of protocols, processes, and voluntarily self-associating networks. Accordingly, the Internet cannot be governed effectively by any one organisation or set of interests. An exclusionary intergovernmental model would be especially ill suited to its unique characteristics; only a truly open, multistakeholder, and flexible approach can ensure the Internet's continued growth and transition into a multilingual medium. In parallel, when the conditions for system stability and sound management can be guaranteed, authority over inherently global resources like the root servers should be transferred to a global, multistakeholder entity.

The international community must have full and easy access to knowledge and information about ICT global governance decision making. This is a baseline prerequisite for implementation of the principles mentioned above, and indeed for the success of the WSIS process itself. We need public-interest oriented monitoring and analysis of the relevant activities of both intergovernmental and "self-governance" bodies including, inter alia, the International Telecommunication Union, the World Trade Organization, the World Intellectual Property Organization, the United Nations Conference on International Trade Law, the Organization for Economic Cooperation and Development, the Hague Conference on International Private Law, the of Europe, the Asia Pacific Economic Cooperation, the North American Free Trade Agreement, the Internet Corporation for Assigned Names and Numbers, and Wassenaar Arrangement.

As a viable first step in this direction, we recommend the establishment of an independent and truly multistakeholder observatory committee to: (1) map and track the most pressing current developments in ICT global governance decision-making; (2) assess and solicit stakeholder input on the conformity of such decision-making with the stated objectives of the WSIS agenda; and (3) report to all stakeholders in the WSIS process on a periodic basis until 2005, at which time a decision could be made on whether to continue or terminate the activity.

3. Conclusion

It is people who primarily form and shape societies, and information and communication societies are no exception. Civil society actors have been key innovators and shapers of the technology, culture and content of information and communication societies, and will continue to be in the future.

Human rights stand at the centre of our vision of the information and communication society[5]. From this standpoint, action plans, implementation,

---

[5] Nothing in this declaration may be interpreted as implying that civil society wishes to engage in any activity or to perform any act aimed at the destruction of any of the rights and freedoms set forth in the International Bill of Rights and other human rights treaties.

financing mechanisms and governance must all be shaped by and evaluated on the basis of their ability to meet life-critical human needs.

Host countries and institutions contributing to and participating in the post-Geneva WSIS process should fully respect the principles enunciated in the Declaration adopted at the Geneva Summit, including those relating to human rights that are fundamental to the information and communications society. These include, but are not limited to the freedoms of expression, association and information.

Toward this end, and in preparation for the second phase of WSIS, an independent commission should be established to review national and international ICT regulations and practices and their compliance with international human rights standards. This commission should also address the potential applications of ICTs to the realization of human rights, such as the right to development, the right to education and the right to a standard of living adequate for the mental and physical health and well-being of the individual and his or her family, including food, housing and medical care.

The full realisation of a just information society requires the full participation of civil society in its conception, implementation, and operation. To this end, we call on all governments involved in the preparatory processes of WSIS to work in good faith with non-governmental and civil society organisations and fully honour the recommendations of Resolution 56/183 of the United Nations General Assembly. In particular, participating governments must honour civil society's right to participate fully in the remaining intergovernmental preparatory processes leading to the second phase of WSIS.

We commit ourselves—independent of the modalities of participation granted to us by governments—to pursuing by all just and honourable means necessary the realization of the vision of the information society presented herein. To this end, civil society organisations will continue to cooperate with one another to develop a Plan of Action for the second phase of WSIS. We call upon the world's leaders to urgently assume the heavy responsibilities they face, in partnership with civil society, to make this vision a reality.

Notes

Part One Introduction

1. The range of titles is truly vast and go from Frank Webster, *Theories of the Information Society* (London, Routledge, 2002), through Manuel Castells, *The Information Age: Economy, Society and Culture* (Oxford, Blackwell, 1996), to Marc U. Porat, *The Information Economy*. (Washington, DC, US Department of Commerce, 1977).

Chapter One

1. For more information, see the US Amnesty International Website. http://www.amnestyusa.org/index.html
2. See Monroe E. Price, *Media and Sovereignty: The Global Information Revolution and Its Challenge to State Power*. (Cambridge: MA., MIT Press, 2002).
3. OECD Information Technology Outlook, *ICTs and the Information Economy 2002*. (OECD, Paris, 2002).
4. Human Development Report 2001, *Making New Technologies Work for Human Development*, p.39. (New York: United Nations Development Programme, 2001).
5. See the Website of the Motion Picture Association of America for more information. http://www.mpaa.org/
6. See, for an elaborate development of this point, Lawrence Lessig, *Free Culture: How Big Media Uses Technology and the Law to Lock Down Culture and Control Creativity*. (New York: Penguin, 2004).
7. See Robert W. McChesney and Dan Schiller, *Political Economy of International Communications: Foundations for the Emerging Global Debate about Media Ownership and Regulation*. (Geneva: United Nations Research Institute for Social Development, Technology, Business and Society Programme Paper no. 11, 2003).

Chapter Two

1. Human Development Report 2001, *Making New Technologies Work for Human Development*, p.8. (New York: United Nations Development Programme, 2001).
2. The resolutions are available on the WSIS Official Website at: http://www.itu.int/wsis/documents/background.asp?theng=en&c_type=res
3. The *Millennium Declaration* is available on the United Nations Website at: http://www.un.org/millennium/Declaration/ares552e.htm
4. Resolution A/RES/56/258 is available on the WSIS Official Website at: http://www.itu.int/wsis/docs/background/resolutions/56-258.pdf
5. WSIS Executive Secretariat, *Proposed Themes for the Summit and Possible Outcomes (WSIS/PC-1/DOC/4-E)*. (Geneva: May 31, 2002). http://www.itu.int/dms_pub/itu-s/md/02/wsispc1/doc/S02-WSISPC1-DOC-0004!!MSW-E.doc
6. See *The Summit: Content, Themes, and Outcome* on the WSIS Official Website at: http://www.itu.int/wsis/newsroom/fact/content_themes_outcome.html
7. Ibid.
8. This illustration was taken from the *WSIS Newsletter N° 1*. (March 2002). The newsletter is available on the WSIS Official Website at: http://www.itu.int/wsis/newsletter/2002/apr/nl_01.pdf
9. Ibid.
10. All of the elements mentioned in this Table will be presented and explained later in the text.
11. This illustration was taken from the *WSIS Brochure 1*. The brochure is available on the WSIS Official Website at: http://www.itu.int/wsis/docs/brochure/wsis.pdf
12. ITU, Press Release: *Global Strategy for the Information Society Takes Successful First Steps* (July 5, 2002). The press release is available on the ITU Website at: http://www.itu.int/newsarchive/press_releases/2002/17.html
13. Declaration A/RES/56/183 is available on the WSIS Official Website at: http://www.itu.int/wsis/docs/background/resolutions/56_183_unga_2002.pdf
14. WSIS Executive Secretariat, *Proposed Themes for the Summit and Possible Outcomes (WSIS/PC-1/DOC/4-E)*. (Geneva: May 31, 2002). http://www.itu.int/dms_pub/itu-s/md/02/wsispc1/doc/S02-WSISPC1-DOC-0004!!MSW-E.doc
15. Civil society associations and private sector entities were accredited in Phase I of the WSIS by the Executive Secretariat. All accreditations for Phase I were automatically extended to Phase II.
16. Daniel Stauffacher, *Report by the Chairman of Subcommittee 1 on Rules of Procedure (WSIS/PC-1/DOC/0009)*. (Geneva: July 4, 2002).

http://www.itu.int/dms_pub/itu-s/md/02/wsispc1/doc/S02-WSISPC1-DOC-0009!!MSW-E.doc

17 The issues raised by the inclusion and actual participation of these actors will be discussed in the second part of this document.

18 Daniel Stauffacher, *Report by the Chairman of Subcommittee 1 on Rules of Procedure (WSIS/PC-1/DOC/0009)*. (Geneva: July 4, 2002).

http://www.itu.int/dms_pub/itu-s/md/02/wsispc1/doc/S02-WSISPC1-DOC-0009!!MSW-E.doc

Chapter Three

1 Kofi Annan, *Secretary-General Describes Emerging Era in Global Affairs with Growing Role for Civil Society Alongside Established Institutions (SG/SM/6638)*. (July 14, 1998).
http://www.un.org/News/Press/docs/1998/19980714.sgsm6638.html

2 Resolution A/RES/56/183 is available on the WSIS Official Website at:
http://www.itu.int/wsis/docs/background/resolutions/56_183_unga_2002.pdf

3 OECD Information Technology Outlook, *ICTs and the Information Economy* 2002, p.18. (Paris: OECD, 2002).

4 Office of the President of the Millennium Assembly, *Reference Document on the Participation of Civil Society in United Nations Conferences and Special Sessions of the General Assembly During the 1990s*. Version 1. (August 2001).
http://www.un.org/ga/president/55/speech/civilsociety1.htm

5 The Charter of the United Nations is available on the United Nations Website at:
http://www.un.org/aboutun/charter/

6 Resolution 1996/31 is available on the United Nations Website at:
http://www.un.org/documents/ecosoc/res/1996/eres1996-31.htm

7 Report A/53/170 is available on the United Nations Website at :
http://www.un.org/documents/ga/docs/53/plenary/a53-170.htm

8 Report A/54/329 is available on the UN Non-Govermental Liaison Service Website at :
http://www.unige.ch/iued/wsis/DOC/341EN.PDF

9 Report A/57/387, page 25. Report A/57/387 is available on the Office of the United Nations High Commissioner for Human Rights Website at:
http://www.unhchr.ch/Huridocda/Huridoca.nsf/0/4b5d557cb16e82b6c1256c3e003933dd/$FILE/N0258326.doc

10 Report A/58/817, page 9. Report A/58/817 is available on the United Nations Website at:
http://www.un.org/dpi/ngosection/N0437641.pdf

11 Kofi Annan, Press Release: *Secretary-General Describes Emerging Era in Global Affairs with Growing Role for Civil Society Alongside Established Institutions (SG/SM/6638)*. (July 14, 1998).
http://www.un.org/News/Press/docs/1998/19980714.sgsm6638.html

12 The Millennium Declaration is available on the United Nations Website at :

http://www.un.org/millennium/declaration/ares552e.pdf
13 Resolution A/RES/56/183 is available on the WSIS Official Website at:
http://www.itu.int/wsis/docs/background/resolutions/56_183_unga_2002.pdf
14 WSIS Executive Secretariat, *Draft Action Plan (WSIS/PCIP/DT/2(Rev.1)-E.* (May 30, 2003).
http://www.itu.int/dms_pub/itu-s/md/03/wsispcip/td/030721/S03-WSISPCIP-030721-TD-GEN-0002!R1!MSW-E.doc
15 WSIS Official Website, *Basic Information about WSIS.*
http://www.itu.int/wsis/basic/about.html
16 WSIS Executive Secretariat, *Draft Declaration based on Discussion in the Working Group of Sub-committee 2 (WSIS/PC-2/DT/2-F).* (February 25, 2003).
http://www.itu.int/dms_pub/itu-s/md/03/wsispc2/td/030217/S03-WSISPC2-030217-TD-GEN-0002!!MSW-E.doc
17 For more information on the private sector at the WSIS, see the Webpage of the Coordinating Committee on Business Interlocutors (CCBI) on the International Chamber of Commerce Website at:
http://www.iccwbo.org/home/e_business/wsis.asp
18 WSIS Official Website, *Basic Information about WSIS.*
http://www.itu.int/wsis/basic/about.html
19 This was taken from a general statement delivered by Ayesha Hassan at the plenary session of PrepCom 1 on behalf of CCBI/ICC. (July, 2 2002).
http://www.itu.int/wsis/docs/pc1/statements_general/ccbi.doc
20 Ibid.
21 Coordinating Committee of Business Interlocutors, *What are the Contents and Themes that Business Supports for the Summit? (WSIS/PC-2/C/0035).* (December 10, 2002).
http://www.itu.int/dms_pub/itu-s/md/03/wsispc2/c/S03-WSISPC2-C-0035!!MSW-E.doc
22 WSIS Official Website, *Basic Information about WSIS.*
http://www.itu.int/wsis/basic/about.html

Chapter Four

1 WSIS Executive Secretariat, *Participation of Private Sector, Civil Society and Other Stakeholders (WSIS/PC-1/DOC/3-E).* (May 31, 2002).
http://www.itu.int/dms_pub/itu-s/md/02/wsispc1/doc/S02-WSISPC1-DOC-0003!!MSW-E.doc
2 The *WSIS Brochure 1* is available on the WSIS Official Website at:
http://www.itu.int/wsis/docs/brochure/wsis.pdf
3 See the Website of the event for more information.
http://www.wemfmedia.org/index.html.

• Notes • 167

4   The *Lyon Declaration* is available on the Website of the event at:
    http://www.cities-lyon.org/en/Declaration
5   The authors of this document are members of the CRIS campaign, but are expressing themselves here as researchers and individuals.
6   For more information on the beginning of civil society participation in the WSIS preparatory process, see the thematic issue entitled "Communication Rights in the Information Society" in *Media Development*, no 4 (2002).
7   See the Voices 21 Website at:
    http://www.comunica.org/v21/.
8   Personal archives.
9   See *Media Development*, no. 4, (2002).
10 A complete report on this meeting is available on the CRIS Website: http://www.crisinfo.org/
11 See Annex II of the document *UNESCO and the World Summit on the Information Society* (Paris: June 2002) at :
    http://www.unige.ch/iued/wsis/DOC/345EN.PDF
12 Civil Society Plenary, *Shaping Information Societies for Human Needs*. Civil Society Declaration to the World Summit on the Information Society. (Geneva: December 8, 2003).
    http://www.itu.int/wsis/docs/geneva/civil-society-declaration.pdf
    The Declaration is reproduced in the Appendix to this book.
13 See the World Summit on the Information Society Civil Society Meeting Point Website at:
    http://www.wsis-cs.org/cs-overview.html
14 Bruce Girard, *Statement on Behalf of the Civil Society Plenary at PrepCom 1 and the Campaign for Communication Rights in the Information Society (CRIS) to the WSIS Press Conference*. (Geneva: July 5, 2002).
    http://www.communities.org.uk/displayResource.cfm?ResourceID=181

Chapter Five

1   Alain Clerc and Louise Lassonde are co-presidents of the Geneva-based NGO *Fondation du devenir* whose mission is to develop thinking on, and concrete applications of, ideas and initiatives that can improve the quality of life for all members of society.
2   The Civil Society Division Website, from which this list was taken, is now closed.
3   WSIS Official Website, *Establishment of a Civil Society Bureau: A Historic Event!* (April 22, 2003).
    http://www.itu.int/wsis/newsletter/2003/apr/a2.html
4   Bruce Girard, [CRIS Info Listserv]. *Civil Society Wins a Place at WSIS Table*. (March 1, 2003).
5   *Civil Society and NGO Open-Ended Bureau Proposal*. (January 30, 2003).
    http://www.worldsummit2003.de/download_en/CS_Bureau_30Jan_ENG.doc

6 Sean Ó Siochru, [CRIS Info listserv]. "Civil Society and NGO Open-ended Bureau" in MEETING: CS Bureau- social movements. (February 13, 2003).
7 Civil society's structures evolved constantly during the WSIS process. This table shows the structures existing as of the end of Phase I.
8 See the Virtual WSIS CS Plenary Group listserv at:
http://mailman.greennet.org.uk/mailman/listinfo/plenary
9 The working groups and caucus listserv addresses are available at:
http://www.wsis-cs.org/caucuses.html

Chapter Six

1 See the press releases available on the Heinrich Böll Foundation Website for more information on this matter at:
http://www.worldsummit2003.org/
2 See Civil Society Volunteers, *Does Input lead to Impact? How Governments treated Civil Society Proposals in Drafting the 21 September 2003 Draft Declaration.*
http://www.worldsummit2003.de/download_en/does_input_lead_to_impact.rtf
3 Alan Toner, *Unzipping the World Summit on the Information Society.* (July 4, 2003).
http://www.metamute.com/look/article.tpl?IdLanguage=1&IdPublication=1&NrIssue=26&NrSection=10&NrArticle=873&ST_max=0&search=search&SearchKeywords=wsis&SearchLevel=0
4 Meryem Marzouki, *Declaration on Behalf of the Civil Society Plenary.* Address to the Intersessional Meeting. (Paris: July 18, 2003).
http://www.iris.sgdg.org/actions/smsi/hr-wsis/hris-cs-180703.html
5 This meeting, convened by the WSIS organizers, was called informal because it had not been foreseen in the official programme of the Preparatory Committee.
6 See the Webpage for the online discussion forum for civil society at:
http://portal.unesco.org/ci/en/ev.php-URL_ID=5444&URL_DO=DO_TOPIC&URL_SECTION=201.html
7 Civil Society Coordinating Group, *Civil Society Statement to PrepCom 2 on Vision, Principles, Themes and Process for WSIS.* (WSISWSIS/CSCG/5). (Geneva: December 18, 2002).
http://www.itu.int/dms_pub/itu-s/md/03/wsispc2/c/S03-WSISPC2-C-0071!!PDF-E.pdf
8 The document is available on-line at :
http://www.ngocongo.org/ngonew/WSIS-CS-ActionPlan-02272003.doc
9 WSIS Executive Secretariat, *Draft Plan of Action (WSIS/PCIP/DT/2(Rév.1)-F).* (May 30, 2003).
http://www.itu.int/dms_pub/itu-s/md/03/wsispcip/td/030721/S03-WSISPCIP-030721-TD-GEN-0002!R1!MSW-E.doc
10 Civil Society Working Group on Content and Themes Drafting Committee, *"Seven Musts": Priority Principles Proposed by Civil Society.* Geneva: February 25, 2003.

• Notes • 169

11 http://www.choike.org/nuevo_eng/informes/996.html
   The document is available on-line at :
   http://www.itu.int/dms_pub/itu-s/md/03/wsispc2/c/S03-WSISPC2-C-0071!!PDF-E.pdf
12 The document is available on-line at :
   http://www.worldsummit2003.de/download_en/WSIS-CS-CT-Paris-071203.rtf
13 ICANN is controlled by the U.S. Department of Commerce, whereas the ITU has always been closed to civil society and rests strongly in the hands of the private sector.
14 Civil Society Content and Themes Group, *Civil Society Essential Benchmarks for WSIS*. (November 2003).
   http://www.worldsummit2003.de/download_en/CS-Essential-Benchmarks-for-WSIS-14-11-03-final.rtf
15 Bruno Jaffré and Jean-Louis Fullsack, *Sommet mondial sur la société de l'information 2003: quand les altermondialistes ratent le rendez-vous*. (February 2004).
   http://www.csdptt.org/IMG/pdf/SMSI2003finsite.pdf.
16 Intervention by Adama Samassékou at the informal meeting on content and themes. (July 2003).
17 This table does not include documents with contributions and reactions to official texts or the "non-papers".
18 Civil Society Plenary, Press Release: *WSIS Process at Prepcom 3*. (September 26, 2003).
   http://www.worldsummit2003.de/en/web/473.htm
19 World Forum on Community Networking, *Mosaïc, no 1*. (July 2003).
   http://www.globalcn.org/fr/article.ntd?id=1552&sort=1.25
20 Bruno Jaffré and Jean-Louis Fullsack, *Sommet mondial sur la société de l'information 2003: quand les altermondialistes ratent le rendez-vous*. (February 2004) . In fact, there were about 100 participants in the Plenary session, out of 481 officially accredited civil society organizations.
   http://www.csdptt.org/IMG/pdf/SMSI2003finsite.pdf.

Chapter Seven

1 This element will become part of a critique of the first phase of the WSIS in Part 3 of this book.
2 See the Digital Solidarity Fund Website:
   http://www.dsf-fsn.org/
3 The statement is avaible on the Heinrich Böll Foundation Website at :
   http://www.worldsummit2003.de/en/web/472.htm
4 WSIS Executive Secretariat, *Draft Declaration of Principles (WSIS/PCIP/DT/1(Rev.1)-E)*. (May 30, 2003).
   http://www.itu.int/dms_pub/itu-s/md/03/wsispcip/td/030721/S03-WSISPCIP-030721-TD-GEN-0001!R1!MSW-E.doc
5 *A Human Rights Portal to the World Summit on the Information Society* can be found on-line at:

http://www.hri.ca/WSIS/

6 The *Statement on Human Rights, Human Dignity and the Information Society* can be found online at:

http://forums.grandlyon.com//files/dhdeclaration_eng.pdf

7 Civil Society Plenary, *Shaping Information Societies for Human Needs.* Civil Society Declaration to the World Summit on the Information Society. (Geneva: December 8, 2003).

http://www.itu.int/wsis/docs/geneva/civil-society-declaration.pdf

8 Civil Society Plenary, Press release: *Civil Society Launches its Declaration at the World Summit on the Information Society.* (Geneva: December 11, 2003).

http://www.thepublicvoice.org/news/2003_cso_declaration.html

9 Human Rights Caucus, *Back to Basics: WSIS and Human Rights.* Address to the Intersessional meeting. (Paris: July 16, 2003).

http://www.iris.sgdg.org/actions/smsi/hr-wsis/hris-speech-160703.pdf

10 Jean D'Arcy (1969). Cited by Cees Hamelink in the Keynote Speech at the Opening Session of the Civil Society Sector Meeting at Prepcom 1.

http://www.geneva2003.ch/home/events/documents/gen_hamelink_en.htm

11 Cees J. Hamelink, "Human Rights in the Information Society." In Bruce Girard and Seán Ó Siochrú, eds., *Communicating in the Information Society.* (Geneva: UNRISD, 2003): 121-163. Available on-line:

http://files.crisinfo.org/cris/hamelink.pdf

12 Kofi Annan, *The Secretary-General's Message for World Telecommunications Day.* (May 17, 2003).

http://www.itu.int/wsis/newsroom/news/telecom/annan.doc

13 See the brochure *Helping the World Communicate* at:

http://www.itu.int/itudoc/gs/promo/gs/wsis/84126.pdf

14 John Barker, [CRIS info listserv]. *Article 19 Critiques Right to Communicate Draft.* (February 5, 2003).

15 See the agenda of the workshop for more information.

http://www.itu.int/wsis/docs/pc2/inf/workshop/flyer4.doc

16 These include the North American Broadcasters Association, Freedom House, and the International Press Institute.

17 ARTICLE 19, Press Release : *ARTICLE 19 Critiques Right to Communicate Draft.* (February 4, 2003).

http://www.iris.sgdg.org/actions/smsi/hr-wsis/list/2002/msg00039.html

18 Cees J. Hamelink, *CRIS Campaign and The Right to Communicate : A Brief Response to ARTICLE 19.* (Geneva: February 24, 2003).

http://www.vecam.org/article.php3?id_article=173

19 Toby Mendel, *The Right to Communicate: An Overview.* Unpublished document. (October 2003)

20 Seán Ó Siochrú, *Democratising Communication Globally: Building a Transnational Advocacy Campaign.* (October 2003).

• Notes • 171

http://sos.comunica.org/docs/dcg.doc
[21] Ibid.
[22] See the Website of the event at:
http://www.communicationrights.org/
[23] For more information, see the Working Group on Patents, Copyrights and Trademarks Website.
http://www.wsis-pct.org/
[24] Civil Society Working Group on Content and Themes Drafting Committee, *Contribution on Common Vision and Key Principles for the Declaration*. (Geneva: February 2003).
http://www.choike.org/nuevo_eng/informes/997.html
[25] Civil Society Working Group on Content and Themes Drafting Committee, *Comments on the Draft Non-paper of the President of the WSIS PrepCom on the Declaration of Principles*. October 30, 2003.
http://www.itu.int/wsis/docs/pc3/president-non-paper/cs_comments_np.pdf
[26] Civil Society Working Group on Content and Themes Drafting Committee. *Contribution on Common Vision and Key Principles for the Declaration*. (Geneva: February 2003).
http://www.choike.org/nuevo_eng/informes/997.html
[27] Civil Society Working Group on Content and Themes Drafting Committee, *Civil Society Priorities Document*. (Paris: July 12, 2003).
http://www.worldsummit2003.de/download_en/WSIS-CS-CT-Paris-071203.rtf
[28] WSIS Executive Secretariat, *Declaration of Principles (WSIS-03/GENEVA/DOC/4-E)*. (Geneva: December 12, 2003).
http://www.itu.int/dms_pub/itu-s/md/03/wsis/doc/S03-WSIS-DOC-0004!!MSW-E.doc
[29] Heike Jensen, "Gender and the WSIS Process: War of the Words." In Heinrich Böll Foundation, ed., *Visions in Process: World Summit on the Information Society, Geneva 2003–Tunis 2005* (Berlin: Heinrich Böll Foundation, 2003): 19-23.
http://www.worldsummit2003.de/download_de/Vision_in_process.pdf
[30] WSIS Gender Caucus, *Input to World Summit on the Information Society*. (Geneva: December 2002)
http://www.genderwsis.org/uploads/media/GenderCaucusInputPrepCom2.pdf
[31] Marc Raboy, "Media and Democratization in the Information Society." In Bruce Girard and Sean Ó Siochru, eds., *Communicating in the Information Society*. (Geneva: UNRISD, 2003): 101-119. Available on-line:
http://files.crisinfo.org/cris/raboy.pdf
[32] Sally Burch, "Global Media Governance: Reflections from the WSIS Experience." In *Media Development*, no 1 (2004).
http://www.wacc.org.uk/wacc/content/pdf/421
[33] Christoph Dietz and Petra Stammen, *The Community Media Forum at the WSIS*. (Berlin: January 27, 2004).
http://www.worldsummit2003.de/en/web/589.htm
[34] The *Draft Plan of Action* is available on the WSIS Official Website at:

http://www.itu.int/dms_pub/itu-s/md/03/wsispc3/doc/S03-WSISPC3-DOC-0003!!MSW-E.doc
35 WSIS Executive Secretariat, *Draft Plan of Action*. *(WSIS03/PC-3/3-E)*. (August 2, 2003).
http://www.itu.int/dms_pub/itu-s/md/03/wsis/doc/S03-WSIS-DOC-0005!!MSW-E.doc
36 Steve Buckley, *Community Media and the Information Society*. (October 2003).
http://www.ifex.org/en/content/view/full/67461/
37 Ibid.
38 Civil Society Plenary, *Shaping Information Societies for Human Needs*. Civil Society Declaration to the World Summit on the Information Society. (Geneva: December 8, 2003). http://www.itu.int/wsis/docs/geneva/civil-society-declaration.pdf
39 Eric Muragana, [WSIS Thetha list serv]. *Statement from the WSIS Civil Society Media Caucus at PrepCom 3*. (September 16, 2003).
http://lists.sn.apc.org/pipermail/wsis/2003-September/000020.html

Chapter Eight

1 Civil Society Plenary, *Civil Society Plenary Statement on Rules of Procedure, Accreditation and Modalities for NGO Participation*. (July 5, 2002).
http://www.iris.sgdg.org/actions/smsi/CS-final-July5.pdf
2 Myriam Horngren, [IS: Voices of the South listserv]. (July 22, 2003).
http://www.dgroups.org/groups/IS/index.cfm?op=dsp_showmsg&listname=IS&msgid=78877&cat_id=2777
3 Meryem Marzouki, *Declaration on Behalf of the Civil Society Plenary*. Address to the Intersessional Meeting. (July 18, 2003).
http://www.iris.sgdg.org/actions/smsi/hr-wsis/hris-cs-180703.html
4 The *Civil Society Priorities Document* is available on-line at:
http://www.worldsummit2003.de/download_en/WSIS-CS-CT-Paris-071203.rtf
5 Verbatim quote from PrepCom President Adama Samassékou's speech to the Civil Society Plenary, Tuesday, September 23, 2003.
http://www.globalcn.org/fr/article.ntd?id=1745&sort=1.10.5.3
6 Arne Hintz, *Civil Society Legitimation is Crucial for WSIS*. (November 2, 2003).
http://www.worldsummit2003.de/en/web/510.htm
7 Civil Society Plenary, Press Release: *WSIS Process at Prepcom 3*. (September 26, 2003).
http://www.worldsummit2003.de/en/web/473.htm
8 Civil Society Plenary, *Civil Society Statement at the End of the Preparatory Process for the World Summit on the Information Society*. (Geneva: November 14, 2003)
http://www.ngocNGOso.org/ngonew/WSIS%20CS-press%20statement.rtf
9 Sally Burch, "Global Media Governance: Reflections from the WSIS Experience." In *Media Development*, no 1 (2004).

http://www.wacc.org.uk/wacc/content/pdf/421
10 Rik Panganiban and Ralf Bendrath, *How Was the Summit? A Helpful List in Case Your Friends (or any Reporters) Ask You.* (December 16, 2003).
http://www.worldsummit2003.de/en/web/577.htm
11 See the Website of the event at:
http://www.wemfmedia.org/
12 See the Website of the event at:
http://topics.developmentgateway.org/ict/
13 WSIS? WE SEIZE! Website.
http://www.geneva03.org/display/about.php
14 Civil Society Plenary, *Civil Society Plenary Denounces WE SEIZE! Repression.* (Geneva: December 12, 2003).
http://www.geneva03.org/display/item_fresh.php?id=29&theng=en
15 Alan Toner, *Unzipping the World Summit on the Information Society.* (July 4, 2003).
http://www.metamute.com/look/article.tpl?IdLanguage=1&IdPublication=1&NrIssue=26&NrSection=10&NrArticle=873&ST_max=0&search=search&SearchKeywords=wsis&SearchLevel=0
16 [Cris-liaison listserv] Private list. (February 13, 2002).

Chapter Nine

1 WSIS Executive Secretariat, *Declaration of Principles (WSIS-03/GENEVA/DOC/4-E).* (Geneva: December 12, 2003).
http://www.itu.int/dms_pub/itu-s/md/03/wsis/doc/S03-WSIS-DOC-0004!!MSW-E.doc
2 The list of contributors are available on the WSIS Official Website.
http://www.itu.int/wsis/funding/contributors1.html
3 Early in the preparatory process, the WSIS Official Website went so far as to affirm that "*we are indeed in the midst of a revolution, perhaps the greatest that humanity has ever experienced*". Originally found at
http://www.itu.int/wsis/basic/about.html
This phrase was later withdrawn from the Website but a Google search in August 2005 showed that it had been cited in no less than 192 documents.
4 *La Tribune de Genève.* (Geneva: September 28, 2003).
5 *Le Temps.* (Geneva: November 15, 2003).
http://www.letemps.ch/dossiers/dossiersarticle.asp?ID=123568
6 Point C9 b of the *Plan of Action* encourages "*the development of domestic legislation that guarantees the independence and plurality of the media*".
http://www.itu.int/dms_pub/itu-s/md/03/wsis/doc/S03-WSIS-DOC-0005!!MSW-E.doc
7 *La Tribune de Genève.* (Geneva: September 28, 2003).

8   WSIS Executive Secretariat, *Plan of Action*. *(WSIS-03/GENEVA/DOC/5-E)*. (December 12, 2003).
    http://www.itu.int/dms_pub/itu-s/md/03/wsis/doc/S03-WSIS-DOC-0005!!MSW-E.doc
9   WSIS Executive Secretariat, *Declaration of Principles*, paragraph 19.
    http://www.itu.int/dms_pub/itu-s/md/03/wsis/doc/S03-WSIS-DOC-0004!!MSW-E.doc
10 Marita Moll and Leslie Regan Shade, "Vision Impossible? The World Summit on the Information Society." In Marita Moll and Leslie Regan Shade, eds., *Seeking Convergence in Policy and Practice*. (Ottawa, Canadian Centre for Policy Alternatives, 2004): 45-80.
11 Steve Buckley, *Community Media and the Information Society*. (October 2003).
    http://www.ifex.org/en/content/view/full/67461/

Chapter Ten

1   Meryem Marzouki, *Declaration on Behalf of the Civil Society Plenary*. Address to the Intersessional Meeting. (July 18, 2003).
    http://www.iris.sgdg.org/actions/smsi/hr-wsis/hris-cs-180703.html
2   See *Does Input Lead to Impact? How Governments Treated Civil Society Proposals in Drafting the 21 September 2003* on the Heinrich Böll Foundation Website at:
    http://www.worldsummit2003.de/download_en/does_input_lead_to_impact.rtf
3   Bruce Girard and Sean Ó Siochru, "Civil Society Embroiled in the System." In *Information Society and International Cooperation: development.com* (Geneva, Institut universitaire d'études du développement, November 2003).
    http://www.unige.ch/iued/wsis/DEVDOT/00299.HTM
4   Sean Ó Siochru, "Will the real WSIS please stand up? The historic encounter of the 'information society' and the 'communication society'." In *Gazette: The International Journal for Communication Studies*, Vol. 66, Nos. 3/4 (June/July 2004): 203-224.
    http://www.crisinfo.org/content/view/full/246
5   Sally Burch, "Global Media Governance: Reflections from the WSIS Experience." In *Media Development*, no 1 (2004).
    http://www.wacc.org.uk/wacc/content/pdf/421
6   Georg C. F. Greve, *Debriefing on the World Summit on the Information Society (WSIS) Geneva Phase*. February 16, 2004.
    http://fsfeurope.org/projects/wsis/debriefing-geneva.html
7   See the on-line Portal Choike:
    http://www.choike.org/nuevo_eng/informes/1955.html
8   Sean Ó Siochru, "Will the Real WSIS Please Stand Up? The Historic Encounter of the 'Information Society' and the 'Communication Society'." In *Gazette: The International Journal for Communication Studies*, Vol. 66, Nos. 3/4 (June/July 2004): 203-224.
    http://www.crisinfo.org/content/view/full/246

Chapter Eleven

1. World Bank, *Managing Development-The Governance Dimension*. (Washington: World Bank, 1994).
2. See the UNDP glossary.
http://magnet.undp.org/Docs/!UN98-21.PDF/!GOVERNA.NCE/!GSHDENG.LIS/Glossary.pdf
3. Claudia Padovani and Arjuna Tuzzi, *Changing Modes of Participation and Communication in an International Political Environment. Looking at the World Summit on the Information Society as a Communicative Process*. Paper presented in the Political Communication section of the IPSA Congress in Durban. (Durban: July 2003).
http://www.ssrc.org/programmes/itic/publications/civsocandgov/Padovani.pdf
4. Ibid.
5. Ibid.
6. Marc Raboy, "The World Summit on the Information Society and its Legacy for Global Governance." In *Gazette: The International Journal for Communication Studies*, vol. 66, nos. 3 -4, (June-July 2004): 225-232.
http://www.er.uqam.ca/nobel/gricis/gpb/pdf_ecrits/Raboy2.pdf
7. Ibid.
8. See Sean MacBride, *Many Voices, One World: Towards a New, More Just and More Efficient World Information and Communication Order*. (Paris : UNESCO, 1980).
9. The ITU defines itself on its Website as "*an international organization within the United Nations system where governments and the private sector coordinate global telecom networks and services.*"
http://www.itu.int/home/index.html
10. Footnote 2 to the Civil Society Declaration.
http://www.worldsummit2003.de/download_en/WSIS-CS-Dec-25-2-04-en.rtf

Chapter Eleven

Bibliography

Documents Produced by Civil Society Within the Framework of the WSIS (by date of publication)

Civil Society Plenary. *Civil Society Plenary Statement on Rules of Procedure, Accreditation and Modalities for NGO Participation.* Geneva: July 5, 2002.
http://www.iris.sgdg.org/actions/smsi/CS-final-July5.pdf

WSIS Gender Caucus. *Input to World Summit on the Information Society.* Geneva: December 2002.
http://www.genderwsis.org/uploads/media/GenderCaucusInputPrepCom2.pdf

Civil Society Coordinating Group. *Civil Society Statement to PrepCom 2 on Vision, Principles, Themes and Process for WSIS. (WSISWSIS/CSCG/5).* Geneva: December 18, 2002.
http://www.itu.int/dms_pub/itu-s/md/03/wsispc2/c/S03-WSISPC2-C-0071!!PDF-E.pdf

Civil Society & NGO Open-ended Bureau Proposal. 30 January 2003.
http://www.worldsummit2003.de/download_en/CS_Bureau_30Jan_ENG.doc

Civil Society Working Group on Content and Themes Drafting Committee. *Contribution on Common Vision and Key Principles for the Declaration.* Geneva: February 2003.
http://www.choike.org/nuevo_eng/informes/997.html

Civil Society Working Group on Content and Themes Drafting Committee. *"Seven Musts": Priority Principles Proposed by Civil Society.* Geneva: February 25, 2003.
http://www.choike.org/nuevo_eng/informes/996.html

Civil Society Working Group on Content and Themes Drafting Committee. *Plan of Action: Civil Society Priorities.* Geneva: February 28, 2003.
http://bscw.fit.fraunhofer.de/pub/bscw.cgi/0/42953741

Civil Society Working Group on Content and Themes Drafting Committee. *Civil Society Priorities Document.* Paris: July 12, 2003.
http://www.worldsummit2003.de/download_en/WSIS-CS-CT-Paris-071203.rtf

Human Rights Caucus. *Back to Basics: WSIS and Human Rights.* Address to the Intersessional Meeting. Paris: July 16, 2003.
http://www.iris.sgdg.org/actions/smsi/hr-wsis/hris-speech-160703.pdf

Meryem Marzouki. *Declaration on Behalf of the Civil Society Plenary.* Address to the Intersessional meeting. Paris: July 18, 2003.
http://www.iris.sgdg.org/actions/smsi/hr-wsis/hris-cs-180703.html

178 • Civil Society, Communication and Global Governance •

Human Rights Caucus. *Statement from Civil Society Organizations on Tunisia and WSIS*. Geneva: September 19, 2003.
http://www.iris.sgdg.org/actions/WSIS/hr-wsis/petition-tunisia-en.html
Natasha Primo (on behalf of the Civil Society Content and Themes Working Group). *Statement of Civil Society in Response to the WSIS Draft Declaration*. Geneva: September 22, 2003.
http://www.worldsummit2003.de/en/web/456.htm
Civil Society Volunteers. *Does Input Lead to Impact? How Governments Treated Civil Society Proposals in Drafting the 21 September 2003 Draft Declaration*. Geneva: September 22, 2003.
http://www.worldsummit2003.de/download_en/does_input_lead_to_impact.rtf
Civil Society Content and Themes Group. *Civil Society Statement on Information and Communication Solidarity Funding Mechanisms*. Geneva: September 25, 2003.
http://www.worldsummit2003.de/en/web/472.htm
Civil Society Plenary. *WSIS Process at Prepcom 3*. Geneva: September 26, 2003.
http://www.worldsummit2003.de/en/web/473.htm
Civil Society Working Group on Content and Themes Drafting Committee. *Comments on the Draft Non-paper of the President of the WSIS PrepCom on the Declaration of Principles*. October 30, 2003.
http://www.itu.int/wsis/docs/pc3/president-non-paper/cs_comments_np.pdf
Civil Society Content and Themes Group. *Civil Society Essential Benchmarks for WSIS*. November 2003.
http://www.worldsummit2003.de/download_en/CS-Essential-Benchmarks-for-WSIS-14-11-03-final.rtf
Civil Society Plenary. *Civil Society Statement at the End of the Preparatory Process for the World Summit on the Information Society*. Geneva: November 14, 2003.
http://www.choike.org/nuevo_eng/informes/1420.html
Civil Society Plenary. *Shaping Information Societies for Human Needs*. Civil Society Declaration to the World Summit on the Information Society. Geneva: December 8, 2003.
http://www.itu.int/wsis/docs/geneva/civil-society-declaration.pdf
Civil Society Plenary. *Civil Society Plenary Denounces WE SEIZE! Repression*. Geneva: December 12, 2003.
http://www.geneva03.org/display/item_fresh.php?id=29&theng=en

Executive Secretariat Documents (by date of publication)

Proposed Themes and Possible Outcomes. (PC-1/DOC/4-E). May 31, 2002.
http://www.itu.int/dms_pub/itu-s/md/02/wsispc1/doc/S02-WSISPC1-DOC-0004!!MSW-E.doc
Participation of Private Sector, Civil Society and Other Stakeholders (WSIS/PC-1/DOC/0003). May 31, 2002.
http://www.itu.int/dms_pub/itu-s/md/02/wsispc1/doc/S02-WSISPC1-DOC-0004!!MSW-E.doc

• Bibliography • 179

Report by the Chairman of Subcommittee 1 on Rules of Procedure (WSIS/PC-1/DOC/0009). July 4, 2002.
http://www.itu.int/dms_pub/itu-s/md/02/wsispc1/doc/S02-WSISPC1-DOC-0009!!MSW-E.doc
Final Report of PrepCom1 (WSIS/PC-1/DOC/0011 (rev. 1)). July12, 2002.
http://www.itu.int/dms_pub/itu-s/md/02/wsispc1/doc/S02-WSISPC1-DOC-0011!R1!MSW-E.doc
Report of the Pan-European Regional Conference for WSIS (WSIS/PC-2/DOC/5-E). January 15, 2003
http://www.itu.int/dms_pub/itu-s/md/03/wsispc2/doc/S03-WSISPC2-DOC-0005!!MSW-E.doc
Report of the African Regional Conference for WSIS (WSIS/PC-2/DOC/4-E). January 15, 2003
http://www.itu.int/dms_pub/itu-s/md/03/wsispc2/doc/S03-WSISPC2-DOC-0004!!MSW-E.doc
Report of the Asia-Pacific Regional Conference for WSIS (WSIS/PC-2/DOC/6-E) January 22, 2003.
http://www.itu.int/dms_pub/itu-s/md/03/wsispc2/doc/S03-WSISPC2-DOC-0006!!MSW-E.doc
Report of the Latin America and Caribbean Regional Conference for WSIS (WSIS/PC-2/DOC/7-E). February 5, 2003.
http://www.itu.int/dms_pub/itu-s/md/03/wsispc2/doc/S03-WSISPC2-DOC-0007!!MSW-E.doc
Report of Western Asia Regional Conference for WSIS (WSIS/PC-2/DOC/8-E). February 5, 2003.
http://www.itu.int/dms_pub/itu-s/md/03/wsispc2/doc/S03-WSISPC2-DOC-0008!!MSW-E.doc
Report on Activities Leading to PrepCom 2. (WSIS/PC-2/10-E). February 10, 2003.
http://www.itu.int/dms_pub/itu-s/md/03/wsispc2/doc/S03-WSISPC2-DOC-0010!!MSW-E.doc
Draft Declaration Based on the Discussion in the Working Group of Subcommittee 2. (WSIS/PC-2/DT/2-E). February 25, 2003.
http://www.itu.int/dms_pub/itu-s/md/03/wsispc2/td/030217/S03-WSISPC2-030217-TD-GEN-0002!!MSW-E.doc
Draft Declaration of Principles (WSIS/PCIP/DT/1(Rev.1)-E). May 30, 2003.
http://www.itu.int/dms_pub/itu-s/md/03/wsispcip/td/030721/S03-SISPCIP-030721-TD-GEN-0001!R1!MSW-E.doc
Draft Plan of Action (WSIS/PCIP/DT/2(Rev.1)-E). May 30, 2003.
http://www.itu.int/dms_pub/itu-s/md/03/wsispcip/td/030721/S03-WSISPCIP-030721-TD-GEN-0002!R1!MSW-E.doc
Declaration of Principles (WSIS-03/GENEVA/DOC/4-E) December 12, 2003.
http://www.itu.int/dms_pub/itu-s/md/03/wsis/doc/S03-WSIS-DOC-0004!!MSW-E.doc
Plan of Action (WSIS-03/GENEVA/DOC/5-E) December 12, 2003.
http://www.itu.int/dms_pub/itu-s/md/03/wsis/doc/S03-WSIS-DOC-0005!!MSW-E.doc

General Analyses of the WSIS

Boulé, Fabrice. "Switzerland and the World Summit on the Information Society: challenges and deficits." In *Information Society and International Cooperation: development.com* Geneva: Institut universitaire d'études du développement, November 2003.
http://www.unige.ch/iued/wsis/DEVDOT/00312.HTM

Burch, Sally. "Global Media Governance: Reflections from the WSIS Experience." In *Media Development*, no 1 (2004).
http://www.wacc.org.uk/wacc/content/pdf/421

Costanza-Chock, Sacha. *WSIS, the Neoliberal Agenda, and Counterproposals*. Presentation for OURmedia III. May 20, 2003.
http://amsterdam.nettime.org/Lists-Archives/nettime-l-0307/msg00051.html

Esterhuysen, Anriette. *Whose "Information Society"? Or: Was WSIS Worth it?* March 23, 2004.
http://www.worldsummit2003.de/en/web/610.htm

Girard, Bruce and Sean Ó Siochru. "Civil Society Embroiled in the System." In *Information Society and International Cooperation: development.com* (Geneva: Institut universitaire d'études du développement, November 2003).
http://www.unige.ch/iued/wsis/DEVDOT/00299.HTM

Greve, Georg C. F. *Debriefing on the World Summit on the Information Society (WSIS) Geneva Phase.* February 16, 2004.
http://fsfeurope.org/projects/wsis/debriefing-geneva.html

Hamelink, Cees J. "The Global Information Society: Visions, People and Power." In *Information Society and International Cooperation: development.com* Geneva: Institut universitaire d'études du développement, November 2003.
http://www.unige.ch/iued/wsis/DEVDOT/00299.HTM

Hintz, Arne. *Civil Society Legitimation is Crucial for WSIS.* November 2, 2003.
http://www.worldsummit2003.de/en/web/510.htm

Human Rights Caucus. *World Summit on the Information Society Recognizes Importance of Human Rights.* Geneva: December 12, 2003.
http://www.ichrdd.ca/english/commdoc/prelease/WSISdec12.html

Khor, Martin. *World Summit on Information Society Skirts Three Key Issues.* December 2003.
http://www.choike.org/nuevo_eng/informes/1516.html

Moll, Marita and Leslie Regan Shade. "Vision Impossible? The World Summit on the Information Society." In Marita Moll and Leslie Regan Shade, eds., *Seeking Convergence in Policy and Practice*. (Ottawa: Canadian Centre for Policy Alternatives, 2004): 45-80.

Ó Siochru, Sean. "Failure and Success at the WSIS: Civil Society's Next Moves." In *Media Action*, no. 254 (March, 2004).
http://www.wacc.org.uk/wacc/content/pdf/964

——. "Will the Real WSIS Please Stand Up? The Historic Encounter of the 'Information Society' and the 'Communication Society'." In *Gazette: The International Journal for Communication Studies*, Vol. 66, Nos. 3/4 (June/July 2004): 203-224.
http://www.crisinfo.org/content/view/full/246

Padovani, Claudia and Arjuna Tuzzi. *Changing Modes of Participation and Communication in an International Political Environment. Looking at the World Summit on the Information Society as a Communicative Process*. Paper presented in the Political Communication section of the IPSA Congress in Durban. Durban: July 2003.
http://www.ssrc.org/programmes/itic/publications/civsocandgov/Padovani.pdf

Panganiban, Rik and Ralf Bendrath. *How Was the Summit? A Helpful List in Case Your Friends (or any Reporters) Ask You*. Geneva/Berlin: December 16, 2003.
http://www.worldsummit2003.de/en/web/577.htm

Raboy, Marc. 2004 "The World Summit on the Information Society and its Legacy for Global Governance." In *Gazette: The International Journal for Communication Studies*, vol. 66, nos. 3-4, (June-July 2004): 225-232.
http://www.er.uqam.ca/nobel/gricis/gpb/pdf_ecrits/Raboy2.pdf

Toner, Alan. *Unzipping the World Summit on the Information Society*. July 4, 2003.
http://www.metamute.com/look/article.tpl?IdThenguage=1&IdPublication=1&NrIssue=2
6&NrSection=10&NrArticle=873&ST_max=0&search=search&SearchKeywords=wsis&Se
archLevel=0

International Organizations

Economic and Social Council (ECOSOC). Resolution 1996/31. July 25, 1996.
http://www.un.org/documents/ecosoc/res/1996/eres1996-31.htm

Joint Inspection Unit (Francesco Mezzathema). *Involvement of Civil Society Organizations Other Than NGOs And The Private Sector In Technical Cooperation Activities: Experiences And Prospects Of The United Nations System (JIU/REP/2002/1)*. Geneva: United Nations, February 2002.
http://www.unsystem.org/jiu/data/reports/2002/en2002_1.pdf

Organization for Economic Co-operation and Development (OECD). OECD Information Technology Outlook. ICTs and the Information Economy. Paris: OECD, 2002.
www.sourceocde.org

Panel of Eminent Persons on United Nations-Civil Society Relations. *We the Peoples: Civil Society, the United Nations and Global Governance (Report A/58/817)*. United Nations: June 11, 2002.
http://www.un.org/dpi/ngosection/N0437641.pdf

United Nations (UN). *United Nations Charter*. 1945.
http://www.un.org/aboutun/charter/

United Nations Development Programme (UNPD). *Glossary of Key Themes*.
http://magnet.undp.org/Docs/!UN98-
21.PDF/!GOVERNA.NCE/!GSHDENG.LIS/Glossary.pdf

———. 2001. Human Development Report 2001. *Making New Technologies Work for Human Development*. New York, Oxford. Oxford University Press 2001.
http://hdr.undp.org/reports/global/2001/en/pdf/completenew.pdf

United Nations Educational, Scientific and Cultural Organization (UNESCO). *UNESCO and the World Summit on the Information Society*. Paris: June 2002.

http://www.unige.ch/iued/wsis/DOC/345EN.PDF
United Nations General Assembly. *United Nations Millennium Declaration (Resolution A/RES/55/2)*. September 8, 2000.
http://www.un.org/millennium/declaration/ares552e.htm

Official Resolutions on the WSIS

International Telecommunication Union (ITU). *Resolution 73*. 1998.
http://www.itu.int/council/wsis/R73.html
International Telecommunication Union (ITU). *Resolution A/RES/56/183*. January 31, 2002.
http://www.itu.int/wsis/docs/background/resolutions/56_183_unga_2002.pdf
International Telecommunication Union (ITU). *ResolutionA/RES/57/238*. January 31, 2003.
http://www.itu.int/wsis/docs/background/resolutions/57-238.pdf

Private Sector Documents

Coordinating Committee of Business Interlocutors (CCBI), *What are the Contents and Themes that Business Supports for the Summit?* December 10, 2002.
http://www.itu.int/dms_pub/itu-s/md/03/wsispc2/c/S03-WSISPC2-C-0035!!MSW-E.doc
Coordinating Committee of Business Interlocutors (CCBI). *The Final Business Statement.* December 12, 2003.
http://businessatwsis.net/mainpages/media/press/news.php?news_id=22&PHPSESSID=1573870a1f64f0427c190fa5860127c6
Global Infrastructure Information Commission. *Declaration Regarding the First Phase of the United Nations World Summit on the Information Society.* December 12, 2003.
http://www.giic.org/events/12-12-03.asp

References on Civil Society and WSIS Issues, Debates and Themes

Annan, Kofi. Press Release: *Secretary-General Describes Emerging Era in Global Affairs with Growing Role for Civil Society Alongside Established Institutions (SG/SM/6638)*. July 14, 1998.
http://www.un.org/News/Press/docs/1998/19980714.sgsm6638.html
———. *Strengthening of the United Nations: An Agenda for Further Change (Report A/57/387)*. United Nations: September 9, 2002.
ARTICLE 19. *Note on the Draft Declaration on the Right to Communicate*. January 2003. Unpublished.
———. *Statement on the Right to Communicate*. Unpublished.
Buckley, Steve. *Community Media and the Information Society*. October 2003.

• Bibliography •

http://www.ifex.org/en/content/view/full/67461/
Castells, Manuel. *The Information Age: Economy, Society and Culture*. Three volumes. Oxford: Blackwell, 1996.
CRIS. *The CRIS Charter*.
http://www.crisinfo.org/content/view/full/98/
——. February 2003. *WSIS PrepCom 2: The CRIS Verdict*.
http://www.wsis.ethz.ch/CRISverdict.pdf
——. October, 2 2003. *WSIS PrepCom 3: The CRIS Verdict*.
http://www.globalcn.org/fr/article.ntd?id=1788&sort=
——. December 8, 2003. *Framing Communication Rights: A Global Overview*.
http://www.crisinfo.org/content/view/full/221/
——. Issue Papers:
Issue 1: *Is the 'Information Society' a Useful Concept for Civil Society?*
http://www.crisinfo.org/content/view/full/171/
Issue 2: *Why Should Intellectual Property Rights Matter to Civil Society?*
http://www.crisinfo.org/content/view/full/176/
Issue 3: *What is the Special Significance of Community Media to Civil Society?*
http://www.crisinfo.org/content/view/full/170/
Issue 4: *Media Ownership: Big Deal?*
http://www.crisinfo.org/content/view/full/172/
Issue 5: *The Corporate Sector and Information Control*
http://www.crisinfo.org/content/view/full/173/
Issue 6: *E-waste: Problems, Possibilities, and the Need for Civil Society Engagement*
http://www.crisinfo.org/content/view/full/169/
Issue 7: *Communities: The Hidden Dimension of ICTs*
http://www.crisinfo.org/content/view/full/174/
Issue 8: *Contesting the Spectrum Allocation Giveaways*
http://www.crisinfo.org/content/view/full/99/
Issue 9: *Universal Access to Telecoms*
http://www.crisinfo.org/content/view/full/175/
Dietz, Christoph and Petra Stammen. *The Community Media Forum at the WSIS*. Berlin: January 27, 2004
http://www.worldsummit2003.de/en/web/589.htm
Girard, Bruce and Sean Ó Siochru, eds., *Communicating in the Information Society*. Geneva: UNRISD, 2003.
http://www.crisinfo.org/content/view/full/231
Hamelink, Cees J. "Human Rights for the Information Society." In Bruce Girard and Sean Ó Siochru, eds., *Communicating in the Information Society*. Geneva: UNRISD, 2003: 121-163.
http://files.crisinfo.org/cris/hamelink.pdf
——. *People's Communication Charter*.
http://www.pccharter.net/charteren.html

———.*Draft Declaration on the Right to Communicate*. Amsterdam/Geneva: December 15, 2002
http://www.ourmedianet.org/documents/wsis_2003/WSIS-CSCG.Declaration_of_CommRights.doc

———.*CRIS Campaign and the Right to Communicate: A Brief Response to Article 19*. February 24, 2003.
http://www.vecam.org/article.php3?id_article=173

———. *Statement on Communication Rights*. December 11, 2003.
http://www.communicationrights.org/statement_en.html

Heinrich Böll Foundation, ed., *Visions in Process: World Summit on the Information Society, Geneva 2003-Tunis 2005*. Berlin: Heinrich Böll Foundation, 2003.
http://www.worldsummit2003.de/download_de/Vision_in_process.pdf

———. *ICANN or ITU? Civil Society Debates Internet Governance*. July 16, 2003.
http://www.worldsummit2003.de/en/web/401.htm

Hurley, Deborah. *Pole Star: Human Rights in the Information Society*. Montreal: Rights & Democracy, 2003.
http://www.ichrdd.ca/english/commdoc/publications/globalization/wsis/polestar.pdf

Le Crosnier, Hervé. *Dans l'œil du cyclone*. September 25, 2003.
http://www.globalcn.org/fr/article.ntd?id=1745&sort=1.10.5.3

Lessig, Lawrence. *Free Culture: How Big Media Uses Technology and the Law to Lock Down Culture and Control Creativity*. New York: Penguin, 2004.

MacBride, Sean. *Many Voices, One World: Towards a New, More Just and More Efficient World Information and Communication Order*. Paris : UNESCO, 1980.

McChesney, Robert W. and Dan Schiller. *Political Economy of International Communications: Foundations for the Emerging Global Debate about Media Ownership and Regulation*. Geneva: United Nations Research Institute for Social Development, Technology, Business and Society Programme Paper no. 11, 2003.

McIver, Jr., William, William F. Birdsall, and Merrilee Rasmussen. *The Internet and the Right to Communicate*. First Monday, Volume 8, Number 12 (December 2003).
http://firstmonday.org/issues/issue8_12/mciver/#m3

Mendel, Toby. October 2003. *The Right to Communicate: An Overview*. Unpublished.

Najar, Ridha. "Voix du tiers monde: Pour un nouvel ordre mondial de l'information." In *UNESCO Courrier*. Paris : December 2001.
http://www.unesco.org/courier/2001_12/fr/medias.htm

Ó Siochru, Sean. *The Case for Mutual Cooperation. A Report to the Secretary-General of the International Telecommunication Union*. October 1995.
http://www.comunica.org/itu_ngo/mutual.doc

———. *Democratizing Communication Globally: Building a Transnational Advocacy Campaign*. 2003.
http://sos.comunica.org/

———. *Transnational Campaign in Media and Communication: What Needs to be Done?* November 13, 2003.
http://sos.comunica.org/

———. *Address to the intergovernmental plenary*. December 12, 2003.

• Bibliography • 185

http://www.itu.int/wsis/geneva/coverage/statements/nexusresearch/c05.doc
Pénélopes (Les). *Radicalement pour le droit à communiquer.*
    http://www.penelopes.org/archives/pages/actualites/agitprop/compa02.htm
Porat, Marc U. *The Information Economy.* Washington, DC: US Department of Commerce, 1977.
Price, Monroe E. *Media and Sovereignty: The Global Information Revolution and Its Challenge to State Power.* Cambridge: MA., MIT Press, 2002.
Raboy, Marc. "Media and Democratization in the Information Society." In Bruce Girard and Sean Ó Siochru, eds., *Communicating in the Information Society.* Geneva: UNRISD, 2003: 101-119.
    http://files.crisinfo.org/cris/raboy.pdf
Webster, Frank. *Theories of the Information Society.* London: Routledge, 2002.
World Forum on Community Networking. *Mosaic: A Synthesis of Civil Society Debates on the Information Society.*
    Mosaic 1. Montreal: July 2003
    http://www.globalcn.org/fr/article.ntd?id=1552&sort=1.25
    Mosaic 2. Montreal: September 2003
    http://www.globalcn.org/en/article.ntd?id=1707&sort=1.25
    Mosaic 3. Montreal: November 2003
    http://www.globalcn.org/en/article.ntd?id=1825&sort=1.25
World Press Freedom Committee. *Statement of Vienna.* Vienna: November 2002.
    http://www.wpfc.org/index.jsp@page=Statement%20of%20Vienna.html
——.*World Press Freedom Committee Response to State Department's Request for Comments on WSIS Documents.* April 2003.
    http://www.state.gov/e/eb/rls/othr/20101.htm

WSIS-related Events

International Symposium on the Information Society, Human Dignity and Human Rights. *Statement on Human Rights, Human Dignity and the Information Society.* Geneva, November 3-4, 2003.
    http://www.iris.sgdg.org/actions/smsi/hr-wsis/list/2002/msg00388.html
World Electronic Media Forum. *Final Report of the World Electronic Media Forum.* WEMF Association Publication. 2004.
    http://www.wemfmedia.org/documents/final_report.pdf
World Summit of Cities and Local Authorities on the Information Society. *Lyon Declaration.* Lyon: December 5, 2003
    http://www.cities-lyon.org/uploadfiles/44.pdf

Internet Websites

Official Websites of International Institutions
ICANN
http://www.icann.org/
Observatory of the Information Society
http://portal.unesco.org/ci/ev.php?URL_ID=7277&URL_DO=DO_TOPIC&URL_SECTION=201&reload=1048272936
United Nations Development Programme (UNDP):
http://www.undp.org
WSIS Official Website
http://www.itu.int/wsis/
UNESCO and WSIS
www.unesco.org/wsis
UN-ICT Task Force
http://www.unicttaskforce.org/index.html
International Telecommunication Union (ITU)
http://www.itu.int/home/index-fr.html
WSIS On-line
http://www.wsis-online.net/

Host Country Websites
Switzerland
http://www.wsisgeneva2003.org/
Tunisia
http://www.WSIStunis2005.org/

Civil Society
Civil Society Meeting Point
http://www.wsis-cs.org/
Civil Society Division Website (now closed)
http://www.wsis2005.org/

Civil Society Caucuses
Africa:
www.wsis-cs.org/africa

Asia-Pacific:
http://www.wsisasia.org/
Cities and Local Authorities:
http://www.cities-lyon.org/
Environment and ICTs WG:
http://www.wsis.ethz.ch/
Gender:
http://www.genderwsis.org/
Human Rights
http://www.iris.sgdg.org/actions/WSIS/
Indigenous Peoples:
http://groups.yahoo.com/group/ip_at_wsis/
NGO Gender Strategies WG:
http://www.genderit.org/
Patents, Copyright and Trademarks WG:
http://www.wsis-pct.org/
Scientific Information WG:
http://www.wsis-if.org/
Trade Unions:
http://www.global-unions.org/wsis.asp
Western Asia and the Middle East:
http://www.irancsos.net/english/index.htm
Youth @ the WSIS:
http://ycdo.takingitglobal.org/wsis

Civil Society Initiatives

AMARC
www.amarc.org
Association for Progressive Communication
http://www.apc.org/english/wsis/
Choike.org
http://www.choike.org/nuevo_eng/informes/703.html
The Daily Summit
http://www.dailysummit.net/
Communication Rights in the Information Society (CRIS)
http://www.crisinfo.org/
Digital Opportunity Channel
http://www.digitalopportunity.org
Electronic Commons / Agora-électronique
http://wsis.ecommons.ca/

Free Software Foundation Europe
http://www.fsfeurope.org/
Global Knowledge Partnership
http://www.globalknowledge.org/
Heinrich Böll Foundation
http://www.worldsummit2003.org/
Human Rights Portal on the Information Society
http://www.hri.ca/WSIS/
ICT for Development Platform
http://www.ict-4d.org/
One World
http://www.oneworld.net/
Swiss Civil Society
http://www.cooperation.net/comunicach/
Taking IT Global
http://www.takingitglobal.org
Voices 21: A Global Movement for People's Voices in Media and Communication in the 21st Century
http://www.comunica.org/v21/statement.htm
WSIS? WE SEIZE!
http://www.geneva03.org
World Association for Christian Communication (WACC)
http://www.wacconline.org.uk
World Forum on Communication Rights
http://www.communicationrights.org/coverage.html
World Forum on Community Networking-Carrefour mondial de l'Internet citoyen (CMIC)
http://www.globalcn.org

Private sector

Business at WSIS
http://businessatwsis.net

WSIS Coverage

Centre des media alternatifs du Québec
http://www.cmaq.net/fr/node.php?id=14625
Journal indépendant du WSIS Terra Viva
http://www.ipsnews.net/focus/tv_society/index.asp

PrepCom 3
http://www.choike.org/nuevo_eng/informes/1339.html#Press%20coverage
Le Temps
http://www.letemps.ch/
La Tribune de Genève
http://www.tdg.ch/tghome.html
WSIS Official Website
http://www.itu.int/wsis/geneva/newsroom/newsletters.asp?theng=en&new=t